Mind Regained

Also by Edward Pols:

The Recognition of Reason

Whitehead's Metaphysics:
A Critical Examination of "Process and Reality"

Meditation on a Prisoner:
Towards Understanding Action and Mind

The Acts of Our Being:
A Reflection on Agency and Responsibility

Radical Realism:
Direct Knowing in Science and Philosophy

Mind
Regained

EDWARD POLS

Cornell University Press / Ithaca and London

First published 1998 by Cornell University Press.

Printed in the United States of America

LIBRARY OF CONGRESS CATALOGING-IN-PUBLICATION DATA

Pols, Edward.
    Mind regained / Edward Pols.
      p.   cm.
    Includes bibliographical references and index.
    ISBN 0-8014-3531-5 (cloth : alk. paper)
    1. Philosophy of mind. 2. Causation. I. Title
  BD418.3.P65   1998
  128'.2—dc21   98-9482

Cornell University Press strives to use environmentally responsible suppliers and materials to the fullest extent possible in the publishing of its books. Such materials include vegetable-based, low-VOC inks and acid-free papers that are either recycled, totally chlorine-free, or partly composed of nonwood fibers.

Cloth printing   10 9 8 7 6 5 4 3 2 1

Contents

Preface

TO SPEAK of mind regained is to imply that mind has been lost. But in this century, in which mind has accomplished such wonders in science and technology, how can it be plausibly implied that mind has been lost? In the course of the twentieth century, the power of mind's functions has revolutionized the study of physics. It has unleashed on this planet the terrible energy that drives the sun and the other stars. It has flung about the planet and indeed about the entire solar system an electronic web so delicate and precise that by virtue of it some of us have walked on the moon, and all of us have been able to inspect, as it were at close range, the moons of Jupiter and the rings of Saturn. That power has constructed and stationed in the clarity of space a telescope whose unexampled reach has opened to our sight an intergalactic vastness far greater than that imagined space whose silence so frightened Pascal. It has constructed a complex blend of intricate theory and instruments that has allowed us to "see" into some of the most fine-grained energy transactions of nature; it has produced another complex of theory and instruments that allows us to "see" (and to aspire to control) the minuscule logical structure of living matter. In all these achievements the power of mind's manifold functions seems only now to be coming into its maturity. In such a time how can it even be suggested that mind has been lost?

My title, I concede, is an exaggeration. But something profoundly important to mind's well-being has indeed been lost, and lost by the very persons who should have been most zealous to preserve it—I mean the most influential workers in academic philosophy, cognitive science, and neurophysiology. Mind, as represented by this group of philosophers and scientists, has failed to see that it operates as a real cause within and upon the material world, and that this real causality is the source of all the theoretical and physical devices that have made possible all the wonders I mentioned. Mind, as represented by that influential body of persons, has lost an adequate understanding of the very functions by virtue of which it accomplishes both its everyday and its more exalted tasks. It has failed in self-knowledge and so has lost resources that are still there to be found.

This failure in self-knowledge on the part of a powerful intellectual establishment has had a negative influence on those educated people who do not belong to that establishment but follow its pronouncements because they find the mind-body problem interesting and important. These members of a wider intellectual community have as a consequence been unwilling or unable to look at their own minds in action and find there what has been left out of that establishment account. This book is aimed in part at that wider public in the hope of bringing about a new self-consciousness on the part of mind as it manifests itself in their persons. In this book I call that kind of intellectual enterprise *attending to mind itself.*

In a preface it is appropriate to speak both roundly and simply about the reasons for the loss I perceive. I single out just the two most important reasons. The first is a failure of philosophy over the centuries rather than a failure in the confluence of philosophy, cognitive science, and neurophysiology I have been calling an establishment. The current of philosophy running from Descartes through Kant and down into contemporary empiricism has predisposed most contemporary philosophers to what I call in this book a negative philosophical judgment about the powers of the human mind. That judgment lays it down that the mind has no direct reality-attaining function but must make do instead with a groping and always-frustrated approach to the real that begins in ideas and then makes its way by constructing theories about that which is not directly accessible.

The second reason is the prevalence in both philosophy and science of what I call the received scientific doctrine of causality. Because of that, the influential and talented writers I describe as an establishment have been determined for much of this century to provide an *explanation* for the mind of which they, the reader, and I are instances. But they are also determined that no explanation of mind shall be regarded as adequate unless it displays mind as an *effect* of the physical entities and events that science investigates so superbly. An acceptable explanation must therefore exclude any *apparent* causality on the part of mind that resists explanation in terms of the doctrine of causality that prevails in science.

Many important things that seem to lie in the province of mind resist explanation in terms of that doctrine. The greatest works of the mind in science, the arts, and morals are pervaded by rational, esthetic, and moral modes of order that make up the very content of those works. The works have the structure they do have in order to make manifest just those particular modes of order. In any particular work—whether of science, art, or morals—its mode of order mandates the particular steps that are necessary to produce the order the person who makes the work is aiming at. We therefore cannot leave reasons (in a sense broad enough to comprise science, art, and morals) out of any really searching causal account of the genesis of such works; and we cannot leave out a mode of causality that mind itself *appears* to exercise in discerning reasons (in this broad sense) and weaving them into the articulation of the work. Despite all this, the establishment resolutely proposes to set aside all such apparent causes as *merely* apparent, and to replace them with causal accounts couched in terms of what I call in this book the received scientific doctrine of causality.

I am sure that these brief remarks on the negative philosophical judgment and on the received scientific doctrine of causality are too compact to be satisfactory, but both themes are developed in a leisurely way in the book itself, and the long introductory chapter that follows this preface provides a clear approach to the program of the book. To that wider educated public I hope to reach I would add this consoling thought: attending to mind itself—the objective of Part Two—is essential to regaining what has been lost, and though attention of that sort is not easy, it is by no means a matter that must be left to professional philosophers and professional students of the mind.

While Cornell University Press was considering the manuscript of this book, three friends, Barbara S. Held, Alicia Juarrero, and Theodore A. Young, read that manuscript and made useful and encouraging comments. Although the argument and structure of the book remain what they were, those comments have led to several local changes that make for greater clarity in what is now before the reader. I am grateful to these friends for their generous help. Held, who is well known for making realism a new and vital force in theory of psychotherapy (1995), brought to her comments (and to our many conversations as this book developed) a sound philosophical perspective, but one from outside the field of academic philosophy, and this has helped me in my effort to make this book accessible to a wider audience. Another friend, Louis Dupré, has played a larger role in the development of the book than perhaps he knows: his enthusiastic and perceptive response to an early version of Part Two helped assure me that there was something useful in what I was trying to say, for he is a person who speaks with authority about philosophical matters. I thank him for that generous support.

Finally, I am happy to say in a public forum of Roger Haydon, editor at Cornell University Press, what I have already said privately: he made it clear to me why an earlier version of this work did not fulfill my aim of writing something

Mind Regained

Introduction

Mind and the Question of Causal Hierarchies

THE STUDY of mind was once an ample and commodious study, drawing into it not only our scientific interests but our moral and religious ones as well. It was long dominated by the conviction that mind itself is the deepest ordering principle of nature or at least the most important expression of that ordering principle. Although I call it the study of mind, it was also the study of the soul and—because soul was often regarded as the life principle—the study of life as well. Sometimes the mind was regarded as the highest factor in the hierarchical structure of the souls of particular human beings. Sometimes a distinction was made between a divine, infinite, and universal Mind and a human, finite, and particular mind; in that case the divine Mind was regarded as the maker of the human soul and thus of the finite mind resident in that soul.

For the West, the conceptual shape of this study of mind was formed by such great philosophical doctrines as those of Plato, Aristotle, Plotinus, and Thomas Aquinas. That doctrinal tradition culminates in the medieval notion of the mind of a creative God, a notion whose power in the West is hard to overstate. For a long time the notion was as important in science as in religion and morals; and although its importance for science gradually diminished after the eighteenth century, it remained a vital force for our culture in general all through the nineteenth century.

Those who still felt the force of that notion were usually ready for some accommodation with science. They might, for instance, concede that a finite mind cannot hear or see or think if certain neural pathways are disrupted, and yet insist that neurophysiological explanations are only part of the story. For such intransigent traditionalists—a numerous band by no means composed only of philosophical idealists—it was just as true that an infinite Mind produced neural structures so finite minds could perform the functions of hearing, seeing, and thinking as that natural selection produced the modern brain and the brain has in turn caused or produced those mental functions.

When traditionalists pressed their case in that way, they were reviving an accommodation that goes back at least to Plato's reaction to the protoscience of his own day. The general movement of our culture, however, suggests that those who believed that Mind is an ultimate explanatory principle were fighting a rear-guard action. The study of mind—at least the study as conducted by the most influential professional philosophers, neurophysiologists, computer scientists, and psychologists—is now anything but ample and commodious.

Those who are likely to read this book are probably familiar with the direction in which the study of mind is now tending. If they have not learned of it in a formal academic setting, they will have read about it in the sophisticated reports of good scientific journalists, the summaries of technical books by influential book reviewers, or even the popularized science and philosophy of the news magazines. The straitened nature of the study in the most dominant academic and public-opinion forums can be seen from the questions that now drive it. Here are a few of them: What is the detailed structure of the central nervous system? How does that total structure work at each discernible level of size, from ions and ion gates up to such relatively larger structures as the amygdala, thalamus, corpus callosum, and hippocampus? How does the central nervous system enable us to be rational? Does the brain, for instance, *cause* consciousness or awareness? Or is the word 'cause' not appropriate for the relation between brain and consciousness? Is the brain perhaps related in some other way to the consciousness it supports? Is it perhaps in fact identical with consciousness, and is the *apparent* difference between consciousness and the brain just what that word suggests—an appearance? Is the brain very like the physical structure of a computer, and is what goes on in our minds very like what goes on in the running of a program on a computer?

It is not my purpose to try to restore the prescientific study of mind just as it was. But the loss of the amplitude of that study—especially its connection with our moral lives—is by no means trivial. Philosophers who moved in the humanistic atmosphere of that traditional study were getting at something that it is important to see and express. Whatever that something is, it is as available to our mental powers as it ever was. If we manage to see it and express it more clearly, I think we shall also see that, despite the brilliant and useful achievements of a study of mind conducted by the methods of the physical sciences, something vital is missing in such a study.

From antiquity until the present certain highly complex entities have seemed so impressive that their functions do not seem to be adequately accounted for if they are regarded as mere assemblages of more basic entities. These complex entities are living beings, but in particular those living beings that are capable of rational action. Rational agents—ourselves, and whatever other creatures, unknown to us except in imagination, may possess that curious combination of agency and mind for which we value ourselves—seem to be entities in some privileged sense rather than entities by courtesy.

The impulse to offer causal explanations of such preeminent beings and their functions is opposed by our tendency to value and respect them. It is not so much that we reject the notion that rational agents can be explained as that we obstinately feel that any explanation that can legitimately be given should not set aside as merely apparent functions what purport to be authentic functions of such creatures—moral choice, for instance. In this book I try to give this obstinate feeling a rational justification: it is my purpose to show that such mind-endowed entities are causes in a sense that we have not yet managed to express, a sense moreover within which we can incorporate and do justice to the kind of causal explanation that science so abundantly provides.

In speaking of causes and entities I have been assuming that we all know what these words mean, and for most of our occasions we do use them with some understanding. But in science and philosophy it is best to be clear in advance about the several ways in which these words have been used in the past. This is especially so at the start of a book in which certain new ways of using such words will be suggested.

In philosophical Greek the usual word for 'cause' is the feminine noun *aitia*. The appearance of *aitia* in philosophy is, however, preceded by its use to attribute responsibility to a person for some act, and this appears to be the root sense of the word. In moral and legal contexts *aitia* usually had a negative sense in its earliest appearances—'charge', 'accusation', 'blame', 'fault'; more rarely it had the positive sense 'credit'. There is a corresponding adjective, *aitios, a, on,* which in the prephilosophical Greek of Homer had the primary sense 'blameworthy', 'culpable' and the secondary sense 'being the cause', 'responsible for'. In Plato's philosophical Greek the neuter noun *aition*, which is formed from that adjective, and the feminine noun *aitia* are used in contexts that call for the word 'cause' in an English translation.

The least we can say about the early use of *aitia* in moral and legal contexts is that a certain entity/being—a rational agent—is singled out as the cause of some event. But by Plato's time an account of the cause or causes of something was a many-sided one, and the word 'cause' therefore had several senses. (Some of these are discussed in Chapter 1.) In the account of the notion of a hierarchy of causes in this introductory chapter, as well as in Chapter 6, I also use the term 'cause' in several senses. Any explanation of mind, if it is a sound account, must also be many-sided, for the notions of cause and explanation are intimately related. A doctrine (or theory) about what kinds of causes there are is usually called a doctrine of causality. Some writers seem to think that the term 'causality' should be reserved for that abstract purpose, and that if we wish to draw attention to the power exercised when something operates as a cause, we should speak of the *causation* exercised by that thing rather than its *causality*. But usage is by no means fixed, and in this book I use 'causality' both abstractly, as when talking about a doctrine, and concretely, as when talking about the power exercised in being a cause.

The equivalent expressions 'an entity' and 'a being' are ubiquitous in commonsense language, in science, and in philosophy. The first expression is probably now the more common one, but there are some good reasons for preferring the second. The Latin noun *ens, entis*, which gives us the word 'entity', seems to have been formed, relatively late, from the Greek *ousia*, which Aristotle uses to refer to particular concrete things—notably human beings. (The Greek *ousia* is a noun formed from the feminine participle of the Greek verb 'to be'.) We thus have every right to use the Anglo-Saxon 'a being' rather than 'an entity' to translate the Greek original. The extensive use of *ens* in medieval Latin—together with the related *essentia* and *esse*—probably accounts for the modern tendency to prefer 'an entity' to 'a being'. But the important standard expression 'human being' gives us at least one good reason for preferring 'a being'; furthermore, 'being', as a general term rather than a designation for a particular such as a human being, is still important in certain traditional kinds of philosophy.

In what follows I continue to use either 'entity' or 'being', and sometimes both, according to the context. I must confess, however, that even when I use 'entity' and 'being' in a more or less commonsense setting, I am conscious of the many high-flown uses philosophers have found for the plain verb 'to be' since the very beginning of philosophy. Some of them are very high-flown

indeed: Thomas Aquinas, for instance, can find no more profound characterization for God than the infinitive of the Latin 'to be'. God, Aquinas tells us, is
esse—that is, his very nature is simply to be. Sometimes he in effect doubles the
use of 'to be', as when, drawing on a medieval tradition, he says that God's
essence is to be—that God's *essentia* is *esse*. Parmenides and Plato are ultimately
responsible for bending the commonplace verb 'to be' in this extraordinary
way, and from their time down to the existentialists who flourished earlier in
this century the resulting philosophical senses of the verb have perplexed, con-
soled, affronted, or amused a great many readers, according to their tempera-
ments and traditions.

Many of these senses, in my judgment, are useful in calling our attention to
things implicit in common sense that science leaves out, but only if it is made
clear just how and why the ordinary sense of the word is being altered. In that
spirit, I make unabashed use of some of those senses in what follows; I hope to
persuade you to take them as seriously as I do.

It is obvious that nature is a hierarchy in at least this sense, that it includes
many relatively large entities (or beings) within which we can find smaller entities that contribute to the structure and functions of the larger ones. Sometimes
the contribution of the smaller entities seems so impressive that we tend to say
that the larger ones are simply composed of the smaller, that they are mere
assemblages of smaller entities. We are especially inclined to do this when the
smaller entities outlast the larger and then enter in due course into the constitution of other large entities: the first large entity is gone, but its ingredients persist and now, in their togetherness, help make up a new but equally transient
large entity.

That is clearly the story of our artifacts, from swords and plowshares to
computers and such massive and complex instruments as particle accelerators.
No matter how complex and ingenious such things are, their origin, their persistence, and their functions depend upon their being assemblages of smaller
entities arranged in a certain order. But it is also the story—or part of the
story—of ourselves as organisms, even though not all of our component entities persist after the death of the body. The number of countable entities in one
of the cells of our bodies is bewildering, and that vast number must be multi-
plied by the number of cells which, as we tend to say, make up that entity we
call a human being.

Let me call such complex entities as human beings and their artifacts *hierarchies of complexity,* leaving open for the moment the question whether they are

also hierarchies in some other sense. Let me also introduce a metaphor for such hierarchies: think of them as pyramids made up of stone blocks, each of which represents an entity. The lowest, or base, level of the pyramid will contain the largest number of blocks; each higher level will contain a smaller number of blocks; the top level will consist of only one block, which I call the apex entity. As with all metaphors, it is important not to take this one literally. In the case of organisms and artifacts, the apex entity is the one that gives an organism or artifact its name—tree, squirrel, man, sword, plowshare, computer—but you cannot identify the apex being by singling it out, as a discrete physical thing, from the rest of the figurative pyramid as you can distinguish the topmost stone from the rest of the stones of a real pyramid.

If the organisms and artifacts have component entities arranged in a certain order, where does that order come from? What is the *cause* of the order? How do we *explain* the order? The answer is not far to seek in the case of our artifacts: our rationality provides the order, and does so in a very thoroughgoing sense. It determines the kinds of entities that must go into the assembly; it provides the plan, or overall form, of the complex entity and it arranges the sequence of steps that is to produce the artifact. Moreover, our rationality persists in the operation or functioning of the artifact: one way or another it is our rationality that puts the artifact—sword, plowshare, computer—to its designed use, that stays with it, so to speak, to sustain its functioning. Even those artifacts we design to be automata, building computers into them to supply them with self-sufficiency, require our constant rational oversight if they are to carry out their functions. Our rationality is thus a causal factor in the production of the artifact, although it is not the only causal factor, since the physical reality of the selected parts and the physical reality of the process of manufacture are also causal factors. What amounts to the same thing, the very existence of our rationality—of our minds—must enter into any explanation of the existence of the artifact.

As for ourselves and living creatures in general, it is not easy to say what provides the order in virtue of which such ultimate physical constituents as molecules, atoms, and yet smaller entities are so arranged as to form organic entities such as cells and organs and these in turn so arranged as to form complex entities such as human beings, other animals, and plants. It was not easy to say when such matters were puzzled over only by philosophers, protoscientists, and theologians; but it is still not easy to say now that legions of authentic scientists have been busy on it for a couple of centuries. Two conflicting accounts of the source of the order have dominated both philosophical and scientific investigations of the problem.

The first causal/explanatory account—first not only in my exposition but also in the order of its appearance in history and in its importance in the development of thought—has been much influenced by the role rationality plays in the design and construction of artifacts. In its various versions, most of which are theological, this account attributes the order of nature in general and the order of organisms in particular either to a divine rational agent (or agents) or to an impersonal source of order (such as Plato's Form of the Good) that is like living minds in the sense that it is sufficient to account for both biological order in general and the rational and moral order displayed by rational agents in particular.

But there are subtle differences between this traditional causal account and the account I have given of the relation between ourselves and our artifacts. We do not manage to impart life to our artifacts as wholes nor to any of the parts out of which we make them; still less do we manage to impart to them the rational and moral order we deploy in the making of them. We remain, that is to say, extrinsic to our artifacts, although we use them, run them, and supervise them. According to some versions of the traditional account, however, the power that produces living entities in general and rational entities in particular remains intrinsic to all the workings of the parts of its products and all the workings of the apex entity itself. The power, to be sure, is understood to be more like the apex entity it eventually gives rise to than like the entities at the base of the pyramid.

The second causal/explanatory account is based on a certain interpretation of the methods and results of science. According to this second account, the order of nature operates primarily in the entities at the base of the pyramid. Whatever functions and operations may appear to belong to entities at higher levels of the pyramid are in fact entirely derivative from the entities of the base level. In a truly searching and rigorous scientific program, the causes of things and thus the explanation of things must be brought back to the base-level entities.

The central point of this scientific account for the study of mind has been expressed in various ingenious ways in the late nineteenth century and in the twentieth century. All these ingenuities have been designed to acknowledge the *apparent* causal significance of the level of rational animals while obdurately referring all *real* causal significance first to the neuronal level and thus (by the very logic of the argument) down to the base level. We need not look at these ingenuities here in this Introduction. Some of them are considered in Chapters 4 and 6.

We cannot go back to one of the versions of the first, or traditional, account, but the general tendency of the first account may nevertheless be closer to the truth than the second one. If we focus our attention on what we actually accomplish with our minds, we may come to see that our doings are not mere appearances of causal autonomy, as the second account claims. We may, that is, come to see that rational animals are not merely hierarchies of complexity but *hierarchies of causality* as well. By that I mean that the being singled out as the apex being—you, me, or some other rational animal—has a causal influence on at least some of the vast number of entities/beings—organs, cells, molecules, atoms, and still smaller things—which from another point of view are merely the apex being's constituents. The same point may also be made in terms of the functions of the apex being: in the case of a human being, any mental function would have a causal influence on what happens within and between the cells of the nervous system, even though that apex function cannot be carried out without the causal support of those neural happenings.

When we identify a cause of an entity, event, or situation, we provide at least part of an explanation of that entity, event, or situation. So if there are causal hierarchies there are also hierarchies of explanations. By this I mean that what goes on in upper levels of a causal hierarchy will provide a partial explanation of what goes on in lower levels. Similarly, what goes on at lower levels will provide a partial explanation of what goes on in upper levels. For example, to explain in fine detail what goes on in the brain when we act rationally, we shall have to know something about what rational action is being performed no less than something about the physical parameters of the nervous system at the start of the action.

It seems clear enough that if there are in fact beings that are apex beings of causal hierarchies, we have not so far managed to express the nature of their existence. But if they do exist, it is clear that they cannot be adequately accommodated within the received scientific doctrine of causality, which is discussed in Chapters 3 and 4. On the other hand, I hope to persuade you in the course of this book that the received doctrine *can* be accommodated in a philosophical account of causality that takes causal hierarchies seriously.

By way of establishing the telos of this book, I close this part of the Introduction with two flat assertions. First: *There is an ordering power intrinsic to nature that cannot be adequately explained in scientific terms.* What we call the laws of nature express important features of that power in a partial and abstract way. The doctrine of causality we have received from the scientific tradition is intimately associated with the laws of nature and so shares in that abstractness. That ordering power works in such entities/beings as subatomic particles, atoms, molecules, cells, organs, and organisms. Its work, moreover, is accomplished not only by virtue of its work in the subatomic particles that popular science tends to call the ultimate building blocks of nature. The ordering power is intrinsic to nature by being intrinsic to each of the entities in my list and more besides. It is thus intrinsic to our rational actions—to our minds—and not just to the subordinate entities without which we cannot be what we are.

The ordering power has, in short, a feature science has tried to exclude from its account of nature from at least the late eighteenth century: a hierarchical feature. The beings in which the power works, making them apex beings of a hierarchy of causes, deserve to be called beings in an exemplary sense. I call them *primary beings,* and I mark their dependence on the common power that is intrinsic to them by calling them *participant powers.*

The claim I have just made comes to you, the reader, as an unsupported affirmation—a dogmatic utterance, an asseveration. You may conceivably believe something of the kind already, but if you do, I have so far added little or nothing to whatever reasons you may have for that belief. My reasons come later. Meanwhile, I supplement that claim with another flat assertion: *The ordering power intrinsic to nature works also in human purpose.* To say so is not to say that nature is guided by a power that is in effect human purpose writ large. My point is rather that our end-directed doings are grounded on an ordering power which, if we could confront it, would account for them in a way more adequate to the lived reality of such doings than the accounts science can provide.

The lived reality of the purposive doings of our minds is familiar enough. From the making of swords and plowshares to the making of computers, from the building of houses to the raising of children and the making of careers, from the making of poems, music, and paintings to the making of doctrines and theories, our mental lives are nothing if not telic. The explanations we give one another for our purposive enterprises are inevitably complex. We do not just mention the goal, or telos, we are pursuing: we mention the means as well, and we articulate the logical connections between the several steps required to reach the goal and between the steps and the end itself. We take the logic itself to be effective; we see that it must be obeyed; we see that it is part of causally producing the desired end; we see that it is part of any adequate explanation of the goal-oriented enterprise.

If we wish to go to the moon, our vehicle must reach an escape velocity determined by the logic of Newtonian mechanics. If we wish to make a career in an enterprise aimed at space exploration, any number of specialized fields lie open to us as goals—metallurgy, rocketry, guidance systems, communications, computer science, aeronautical engineering—provided only that we are willing to take on the particular telic exactions of whatever field we choose. We make our general purposes effective by making the subordinate purposes effective.

The inclusion of our purposes in the explanation of our actions is nowhere more important than in our moral lives. It is important—at least we know that it *ought* to be important—to be moved to action by a good end rather than a bad or even a merely trivial one. It is important also that the end should *move* us. It is important, that is to say, that the end—and a good one at that—should become part of the *power* by virtue of which we act. If we try to circumvent this point by giving a thoroughgoing scientific causal account of what purports to be an end-directed moral act, the moral nature of the act simply vanishes. It is not that the scientific account is not relevant: it is just as relevant as the account of any organism in terms of the atoms of which it is composed. But, like that one, it is a partial account.

   The conscious pursuit of ends by human beings is now widely interpreted as an adaptive device of merely local significance—a highly successful outcome of natural selection and so based on chance rather than on an ordering power of which our telic drives, whether merely practical or also moral, are imperfect expressions. On this interpretation, so-called telic phenomena are precisely that—phenomena. They are appearances generated by a reality that has nothing either good or mindlike about it. My flat assertion contradicts this interpretation. It maintains instead that the causal—hence explanatory—value we attribute to our telic drives and their goals is real rather than merely apparent.

Two Approaches to the Study of Mind

   It is obvious that one reason for the narrowing of the study of mind is the exponential growth of science in the last two hundred years. But there is another and less obvious reason. As it happens, that reason is of some importance for the thesis of this book, so I want to prepare the way for it by distinguishing two approaches to the study of the mind. I give them arbitrary names here; my account of the two approaches will, I trust, justify the names. The first approach is by way of *mind itself;* the second by way of the *infrastructure of the mind.*

   It is easy enough to say what I mean by the approach to the study of mind by way of its infrastructure. In the narrowest sense, the study of the brain by the best methods available to science is the study of the infrastructure of mind. In a broader sense, the scientific study of the body in general is the study of the infrastructure of the mind. In a still broader sense, the scientific study of every physical influence that bears upon the body, from nutrition to the environment in general, is the study of the infrastructure.

   By using the expression 'infrastructure of the mind' I am by no means denying the importance of the infrastructure. As a causal *factor* it is of massive importance in all the doings of embodied mind. In this century we all take it for granted that if the infrastructure is compromised, mind itself cannot do what it normally does. If, for instance, some failure in the vascular system cuts off the blood supply to certain areas of the midbrain and brainstem, then short-term memory, and with it the power to concentrate on what is immediately present, is diminished. By assigning such things as the brain to the infrastructure of mind, I do not intend to dismiss that massive causal contribution: this book is about *embodied* mind. I am merely claiming that the physical basis of mind is not the only causal factor in mind.

Still, the distinction between mind itself and its infrastructure is not easy to justify. By insisting on the notion of infrastructure I have in effect said that something profoundly important has been left out of that approach to the study of mind, something so important that I have chosen to call the missing thing *mind itself.* For those committed to what I have called the narrow version of the study of mind, nothing of causal significance is missing when the study of mind is confined to what I call the infrastructure. They would therefore argue that it is a mistake to say that such things as the central nervous system belong to the infrastructure of mind, for no truly *causal* factor is missing when the study of mind is approached only by way of the central nervous system. And they may go on to insist that if that point is not yet clear to intransigent traditionalists, it will become clear to them when the research program now under way is sufficiently advanced.

These imagined objections should make it clear that I am entitled to the expression 'infrastructure of the mind' only if I am also entitled to the expression 'mind itself'. To put the matter another way: if the approach by way of mind itself reveals causal aspects of mind that are missed when mind is approached only by studying the brain, then it is appropriate to say that the brain belongs to the infrastructure of mind; otherwise, it is not appropriate. In that sense the notion of the infrastructure of the mind implies that rational animals are causal hierarchies.

The approach by way of mind itself is in some respects close to common sense, but it is nonetheless elusive and difficult to characterize. It may help if I offer some other names for what I mean by mind itself. I mean the *full concreteness,* the *full actuality,* the *wholeness* of mind, the *lived reality* of mind. The central nervous system can be singled out from the human being who speaks, argues, chooses, feels, and all the rest, and it can then be studied just as a complex biological entity. Although the central nervous system is a thing of flesh and blood and so concrete enough, it is then being considered in abstraction from the full concreteness of mind itself.

It is tempting to try to draw your attention to mind itself by bringing in philosophical jargon. Mind itself, for instance, seems to be one *functional level,* and the level in which we distinguish the hippocampus from the corpus callosum seems to be a different functional level. The level at which we distinguish billions of neurons seems to be yet another functional level. We might then go on to describe these functional levels as *levels of reality;* or we might instead draw on a Greek rather than a Latin root and call the levels *ontological levels.* Professional jargon of that kind has its uses, and as one who has had recourse to it often enough, I have no right to dismiss it. But the more terminology we import, the more we are likely to fail to notice that mind itself is something familiar, something that is part of our commonsense grasp of ourselves in our world.

To bring mind itself into view, you need only *do* something in the rational line and, so to speak, watch yourself doing it. Alternatively, you can watch someone else doing something in the rational line. In short, you can focus on rational action, either your own or someone else's. You see mind itself at work whenever you listen to some friend talking fluently and in an insightful way, whenever you watch someone solving a problem in mathematics or logic on a chalkboard, writing the steps down and explaining the steps orally even while writing.

You can also see mind itself at work by invoking the imaginative powers of your own mind as you read a biography or a letter. The biography or letter is of course also an instance of the work of mind itself, but it is the content of the letter or biography that I want to bring out. Reading a biography of Mozart or some of his letters, for instance, you can "see" Mozart playing solitary billiards, as he often did. No doubt playing billiards is not a major rational action. Still, it is an exercise in hand and eye coordination, and one that may fairly be called a practical exercise in Newtonian mechanics, which in those very days was being brought to perfection by another instance of mind itself in action—Laplace's work over some decades to bring Newton's great project to completion. (I consider Laplace's work in Chapter 3.)

Mozart, however, may sometimes have been busy about something else all the while he played billiards by himself—*mit Herrn Mozart*—as he says in a letter to his wife. He may sometimes have composed "in his head," interrupting his game now and then to go over to his standing-desk and write something down in his score. And though Constanze could have watched him if she were there, and no doubt often did, she could also watch him "in her head" as she read his letter. I watched him "in my head" as I read that same letter, and you watch him so now as you read these pages.

Rational action is probably the highest expression of mind itself, although not many rational actions have the amplitude and depth of those of Newton, Mozart, and Laplace. Mind at work in rational action has a feature that is at once familiar and baffling: an utterly inimitable combination of an inward and an outer manifestation. This is most obvious in those rational actions that take a firm hold on the physical world—actions like playing billiards, playing a piano, writing symbols on a score or a chalkboard, inventing and constructing a computer, operating a computer, writing a letter, conversing with a friend, sawing a piece of wood, putting a log on the fire. All these doings make changes in the physical world: they use it, deploy it, change it, bend it to their own purposes.

Yet such actions are guided by an inwardness that is both active and receptive. (You and I exemplified one active feature of that inwardness when we imagined Mozart engaged in his complex action.) We call that inwardness consciousness, subjectivity, or awareness; but those names tend to overemphasize the receptivity of inwardness—the taking in of what is going on, together with the feeling that accompanies the taking in. There are, of course, occasions on which inwardness is almost entirely receptive, as when a partially sedated person awaits what a surgeon or dentist is about to do. But in rational actions of the kind we have been considering, the agent in question is also intending or purposing to bring something about. The agent can also be receptively aware of the telic drive that is intrinsic to the act, but that telic drive itself is nothing if not active.

The tight union of inwardness and outwardness in the performing of rational actions is inimitable. Although it is utterly familiar, there is nothing else in nature we can compare it to. The union is so intimate that it does not in the least suggest that disastrous dualistic image which is so common in philosophy since Descartes—consciousness and the physical, each in principle complete without the other. The inwardness of rational action is an awareness of outwardness and an awareness also of inwardness's capacity to manage the outward. It is an awareness, moreover, that the capacity to manipulate the outward is a capacity of *mind's* inwardness. Indeed, most of the business of mind's inwardness is with that which is outward. It may well be that the first business of mind's inwardness is to acknowledge the presence of outwardness and to attend to what that outwardness is.

That business purports to be a *causal* business: inwardness in the guise of consciously directed purpose purports to get outward things done. It purports to make the balls on the table go through the movements prescribed by whatever version of billiards is being played. It purports to have brought about the musical composition while, say, keeping to the general rules of the sonata form. It purports to write *something* on the chalkboard, whether or not that something solves the problem in question. It purports to have solved the problem if the problem is indeed solved. It purports to have cut the piece of wood, with whatever degree of precision the available tools afford. And if we take *mind itself* seriously—if, that is, we do not convince ourselves in advance that we must find some deeper account of such doings, one in which the effectiveness of mind's inwardness is represented entirely as a mere appearance of causal effectiveness—then it really does what it purports to do.

The subject matter of the study of mind itself is nothing if not familiar to us. It is concretely and actually present to us as rational animals who deploy our rationality in the world. Mind itself so dominates our existence that we are familiar with it in a way we cannot be familiar with items of the infrastructure that support its deployment. Some such items—neurons, for instance—are also accessible for study, but they are not accessible *as supporting mind*. It is only by deploying mind itself in the theoretical activity we call science that we can learn about that support. Other infrastructure items—electrons, for instance— are so different in scale from ourselves that they are accessible only by way of theory. Theories about them, mathematical models of them, empirical evidences of their presence stand in for them, and we only know *that* electrons are and *what* they are by virtue of these stand-ins. Yet we know well how to *use* neurons and so also electrons, even though we can give no account of that "how": we need only deploy mind itself in whatever task or problem happens to interest us. If the infrastructure is healthy, it will support that activity of mind itself. Indeed our very mental activity (so theory and experiment tell us) brings about new developments in the infrastructure. Learn a new language, devote yourself intensively to music, and changes that further such activity will take place in your brain. Yet in the doing of the activity only mind itself is manifest to us.

Our familiarity with mind itself, however, is a curious one. We do not experience mind itself as we experience (by virtue of mind itself) a color, a smell, or an ordinary physical object like a chair. Our familiarity with mind itself is by way of mind's doings: we must perform one of mind's typical functions in order to be familiar with that function. Our familiarity with mind itself, in short, is reflexive. We may be unable to give an adequate account of what reflexiveness is, and probably we shall never be in a position to say how it is possible—how, for instance, some infrastructure items might subserve reflexiveness. Yet reflexiveness, by virtue of which we are familiar with mind itself, is also familiar to us: we need only perform some rational action and attend to ourselves performing it. Reflexiveness, then, is a vital feature of what I earlier called the inwardness of mind itself.

Many professional specialists in what is often called philosophy of mind make much of a distinction between the macroscopic properties of physical things and their microscopic properties, and they often extend this distinction to mind as well. Consciousness, for instance, is sometimes called macroscopic in contrast with the neurons and the components of neurons, which are called microscopic. In this atmosphere it is tempting to make the point that mind itself is something macroscopic in contrast with microscopic infrastructure elements. But the term 'macroscopic' is not really satisfactory for mind itself.

We can call the brain macroscopic in comparison with its neurons; we can call the neurons macroscopic in comparison with the minuscule ion gates that subserve the electromagnetic activity of neurons. But the mind itself does not have a certain size, and so we cannot call it macroscopic in comparison with either neurons or ion gates. We can, to be sure, say that the mind is a feature of the total organism and that the total organism is macroscopic as compared with its smallest physical parts. But it seems best to avoid the term 'macroscopic' when we are talking about mind itself, for mind does not have physical parts, even though billions of neurons contribute to its workings.

The approach to the study of mind by way of mind itself does, however, allow us to distinguish several kinds of work the mind performs, for instance, perceiving, conceiving, imagining, and drawing logical conclusions. Such kinds of work are usually called functions, but the terms 'work' and 'function' have roughly the same meaning. In turning Aristotle's Greek into English, for example, translators often use the word 'function' where Aristotle has written *ergon*, which is the common Greek word for work. (Recall also that the root sense of the Latin-based 'function' is performing or doing.) This brings us back to the topic of rational action. Rational action seems to be a master function within which we can discern other functions brought together under the telos that defines the action itself. Mozart's complex and extended action of composing as he plays billiards brings together in the unity of one telic drive the functions I mentioned above—perceiving, conceiving, imagining, drawing conclusions—as well as the many other functions I discuss in Chapters 5 and 6.

Rational action, moreover, is characterized by a resistance to any analytic effort designed to represent it as entirely an effect of causes other than itself. If you try to parse an action into a series of events (whether physical or mental) in which any particular event is caused by an earlier event or events, the unity of the act reasserts itself. The very wholeness of the act is violated by such linear-event analysis. The unit of action is not temporally linear: it requires a certain amount of time to be the act it is, and that unit of time is global: the earlier "parts" anticipate the later "parts"; the later "parts" retain the earlier "parts" in order to complete what was begun there. We see this unity most clearly in such acts as the speaking or writing of a sentence or the inventing or performing of a musical phrase. Linear-event analysis clearly breaks down in such cases: the first word or note is vital to the last word or note, but it is not vital by way of *causing* the second word or note, which then causes the third, and so on to the end.

If, on the other hand, you try to parse the rational action into the discrete causal contributions made to it by units of its infrastructure, once again the action resists a complete causal analysis. The causal contributions of discrete infrastructure elements are no doubt real enough, but they do not appear as such in the act. They seem rather to be *used* by the act. But in being used they are absorbed into its temporal unity: to find them one must focus on something different from the act. And to do that one must perform other acts—the theoretical, analytical, and empirical acts that go into neurophysiological studies and studies in basic physics. With respect to its infrastructure, then, the rational act seems to be self-caused in the limited sense that it makes use of the units of the infrastructure by incorporating them into its own actuality. Much of that actuality, moreover, consists of what mind comes to know in the course of the development of the act—the reasons it understands and assents to, the things it understands to be good and therefore to be pursued, the things it takes to be bad and therefore to be avoided. Things thus known are causes of the action in the limited sense that knowing is part of the action and these are things known.

Close as it is to common sense, mind itself is truly elusive. Two of its most elusive features, moreover, seem to pull us in different directions. On the one hand it seems that mind itself is (for you) precisely *your* mind. To put it another way: your mind itself is something you can attend to more adequately than I can; my mind itself is something I can attend to more adequately than you can. Although both of us can attend to the outward component of the mind itself of the other, neither of us can attend to the inward component of the mind itself of the other.

But mind itself eludes us in another direction, for when you look for your mind itself, you are not looking for what defines you as just the particular person you are. You are looking also for something that seems to include standards that are not in the least particular to you. This is a complex matter, and here at the beginning I only touch on what is most obvious: if you attend to what purports to be your mind itself and you find there the conviction that seven plus five equals twelve when you want it to equal twelve, and that it equals eleven when you want it to equal eleven, then you may be attending to *yourself,* but you are not attending to your *mind* itself. With the reflexivity of mind, then, comes the odd fact that mind is not merely individual; it is also universal or general: its presence excludes utter exclusiveness and privacy.

Doctrines and Actuality

As I said earlier, the exponential growth of science in the last two hundred years is only one reason for the narrowing of the study of the mind. The other and less obvious reason is the negative philosophical judgment about the powers of the mind that came out of the long era of philosophical reflection on the nature of knowledge that began in the early work of Descartes and ended in the work of Hume and in Kant's response to Hume—roughly from 1619 to 1790. That negative judgment comes down to us as a philosophical *doctrine*, and the fascination with which we focus on the doctrine stands in the way of our access to mind itself.

The story of that great age of theory of knowledge, or epistemology, has often been told. In barest outline it is this: the premises of Descartes's particular version of rationalism generate not just other rationalisms but also, beginning with Locke, what we now call empiricism. Hume's developed version of empiricism then seems to demonstrate that we can *know* neither the commonsense world nor the causal connection that common sense supposes itself to find within that world; we must instead settle for mere *belief* in those things. In effect that belief is alleged to be a kind of construction by ourselves, as "knowers," of things we had taken to be independent of our own cognitive activity.

Science, the very activity that early empiricism had set out to glorify and indeed to augment by developing a science of human nature itself, now seemed threatened by this verdict of developed empiricism. How could that great paradigm of knowledge, the achievement of Newton, continue to pass as knowledge? Kant, to save the cognitive status of science, propounds a fatal compromise. Yes, we shall continue to be able to say that we *know* the commonsense world, that we *know* the causal connections that exist within it, that we *know* the great body of Newtonian theory to be true. But we can do so only by conceding that all our knowledge is of appearances, or phenomena. Expressed in other terms, our commonsense and scientific knowledge is the result of a constitutive or formative activity on the part of human understanding and sensibility. As Kant's own most compressed statement of his position has it, objectivity is subjective in origin (Kant 1781/1787/1929, 25 [Bxxiii]).

Despite the counterrevolution of nineteenth-century idealism—that convoluted attempt to seat the unaided reason on a throne more secure than even the great rationalists dreamed of—the theories of knowledge that dominate twentieth-century philosophy descend more or less directly either from Hume or from Kant. Sometimes they oscillate between the doctrines of those two figures; sometimes individual philosophers rehearse in their own careers the movement from Hume to Kant.

Although in this book I do not attempt to retell the story of theory of knowledge from Descartes to the present in any detail, I do discuss, in Chapters 2, 3, 4, and 5, significant themes from it that are related to my main argument. For the purposes of this Introduction, I would make only one point about this long and intricate development. The philosophers who carried through these studies in theory of knowledge supposed themselves to be studying what they called the understanding, the reason, or the mind; and their approach to the study was by way of what I have called mind itself rather than by way of the infrastructure of mind. Indeed, until the nineteenth century very little was known about such matters as neurophysiology, and so little could be said about the infrastructure. Speculation there was aplenty: we find it as early as Descartes's observations about what he called the animal spirits; but of effective science there was little enough.

In short, the negative philosophical judgment about the powers of the human mind was reached by the study of mind itself—a study conducted of course by mind itself. I call the judgment negative because its most profound claim is that the mind cannot get at reality, at what is the case, at what is independent of mind's own capacity to believe or to construct. It needs to be emphasized that the negative judgment still hovers over science itself, even as it did in the days of Hume and Kant. If that judgment is taken seriously, science cannot be exempt from it. An incapacity on the part of the mind to know what is real in independence of the mind's own constructions is a general incapacity. It is thus a paradox that the negative judgment has turned attention away from the study of the mind itself to the scientific study of the infrastructure of mind. There are, to be sure, many ingenious ways of making an exception in favor of the mind's activity in science: the realist-antirealist controversy that dominated philosophy of science in the second half of the twentieth century provides many examples of such ingenuities.

In my view this negative philosophical judgment is simply wrong, incorrect, not true; and I have argued so in some detail (Pols 1992). But here at the beginning of the present book I want to make a related point that can be made more simply. It is this: doctrines—together with theories, which we may for the moment think of as doctrines of a certain kind—are the result of the intercourse of our minds with the actual world and with other doctrines. Sometimes doctrines are inadequate to what they purport to be about, sometimes more or less adequate. But either way, they can so command our attention that we are totally unable to turn our minds to the actual world.

In this case, the actual world we wish to study includes the mind itself, and a doctrine to the effect that the mind cannot get at *any* actuality, let alone the highly subtle actuality which is the mind doing its work, stands in our way. One way to deal with the negative judgment is abrupt enough: set the judgment aside; see it for what it is: a simple philosophical mistake transformed into the myth of the self-enclosure of mind. But recall that the mistake emerged out of an earlier approach to the study of mind by way of what I have called mind itself. The abrupt gesture therefore carries a positive task with it: we must take up that approach again; we must entrust ourselves to mind's reflexive capacity to attend to its own functions even as they carry out their tasks in the world in which we find ourselves. I suggest that if our attention is adequate, we shall see what the functions actually achieve, and we shall also see that they cannot be adequately accounted for in terms of the dominant scientific doctrine of causality.

The Organization of This Book

Although the positive argument of this book is directed towards the *actuality* of mind rather than doctrines *about* mind, I begin with a further consideration of certain doctrines I have touched on in this introductory chapter. Part One is accordingly titled "Attending to Doctrines." In Chapter 1, I consider the two most important sources of the notion that causality is hierarchical and that mind is central to that hierarchy. Both of those sources, Plato and Aristotle, take it for granted that there are things that are beyond the power of mind; but they also take it for granted that mind's reality-attaining powers are considerable. They approach the study of mind by way of mind itself rather than by way of its infrastructure. Nonetheless, they acknowledge, although in their own terminology, that the human mind does indeed have an infrastructure.

Chapter 2 is devoted to Descartes's heroic effort to renew the approach to the study of mind (and of the world as well) by way of mind itself. The chapter is a condensed report of those features of his doctrine which in my view make that effort a heroic failure—a failure in the sense that the doctrine carries within itself the seeds of the negative philosophical judgment about the powers of the mind.

In Chapter 3, I deal in more detail with the origins and nature of the received scientific doctrine of causality; in Chapter 4, I consider the limited applicability of that doctrine to the study of mind and suggest a view of the laws of nature more hospitable to the claim that there are causal hierarchies in nature. Both of these chapters depend heavily on the background provided by Chapters 1 and 2.

In Part Two, "Attending to Mind Itself," I have done my best to focus on the actuality of mind rather than on doctrine and theories. On the other hand, any

PART ONE

Attending to Doctrines

1

Plato and Aristotle on Mind, Soul, and Causality

Plato's Forms

THE CONTROLLING feature of Plato's doctrine of mind and soul—and indeed of his philosophy in general—is his insistence that the term 'being' should be reserved for the Forms, or Ideas. (I capitalize those terms when I intend them in Plato's sense.) The Greek for 'form' is *eidos;* the Greek for 'idea' is of course the source of our own word, and it is spelled in the same way. In what follows I usually speak of the Platonic Forms rather than Ideas for three reasons: first, 'form' is the commoner of the two terms in Plato's writings; second, the term 'idea' is often used in modern European philosophy—notably in Descartes's writings—to describe something that exists in minds and nowhere else, which is not at all the status Plato ascribes to the Forms; and, finally, the term 'form' (*eidos*) is common to both Plato and Aristotle, although there are important differences between them about the status of form.

For Plato the Forms are *what* mind knows *when* it knows, but they are independent of the mind—they transcend the mind as well as all particular things—and so are neither ideas/concepts (in the modern sense) nor mere common (general, universal) terms that apply to many particular things. In Plato's view, the Forms are not abstracted from particular things: they simply *are* in their own right, and, far from being dependent on the world of physical things, they provide, according to Plato, the being in which physical things merely participate. In all the developmental stages of a long philosophical life,

Plato insists that all physical (bodily) things come into being and then pass away into not-being, and that we should therefore not say that the things of that world *are* (have being) but rather that they *become*. The word 'become', containing as it does the verb 'to be', does not translate precisely what Plato is saying, for he wants to contrast changing things with the things that *are*. The Greek noun usually translated by 'becoming' is *genesis*, which means birth or generation; Plato uses both that noun and related verbs to speak of things that change. But the word 'becoming' is so much the standard translation that it would be unwise to resort to a different English word.

It is often said in English that Plato stratifies all *reality* into Being, becoming, and not-being, but we do not do justice to Plato's position unless we keep it in mind that where modern writers use the Latin-based term 'reality' (literally, thinghood) Greek philosophers would have used either the Greek word for being or else some other form of the verb 'to be'. If we now substitute the word 'being' for 'reality' in speaking of Plato's stratification, we find ourselves saying something odd: that Plato stratifies all being into Being, becoming, and not-being. It is difficult not to fall into that or a similar verbal oddity, because the various forms of the verb 'to be' are everywhere in our language. There are places in the late Plato where he himself produces such oddities: for instance, in various places in *Parmenides* and also in *Sophist* (251C–252E) he acknowledges a Form 'being', which duplicates the name (Being) of the realm of Forms in which it resides.

As for the multiplicity of the particular, changing things that are available to us by way of our senses, the most we can say about them in Platonic terms is that they *participate* in the Forms—participate, at any rate, while they are coming-to-be and then, in due course, passing into not-being—and that this participation is a kind of dependence. The verbal oddity appears again in the phrase 'coming to be', for only the Forms can be. The changing physical world we know in commonsense terms is suspended somewhere between Being and not-being; in Plato's terms, it is not the object of knowing but rather of belief or opinion.

Plato seems to distrust an inflexible technical terminology, for he always gives us an abundance of alternative terms. In place of 'form' or 'idea', for instance, he sometimes uses the Greek expression *auto ho esti*, which comprises 'itself' and 'it is', together with the name of a Form. In the case of the Form 'beauty', this complex Greek expression usually moves translators to such English translations as 'absolute beauty', 'essence of beauty', and 'beauty itself'. In Plato's later work, certain Forms which he then supposed to be of especial importance for the relatedness among the Forms—sameness and difference (otherness), motion and rest, and (as already noted) being—are called *genē*, which is usually translated as 'classes'. Another important Platonic word for a Form, *paradeigma*, is sometimes translated as 'model' and sometimes taken over directly into English as 'paradigm'. That last word is now commonly used for a somewhat different purpose in philosophy of science, and it is so bandied about in other fields that I suspect some of its users suppose it to be a twentieth-century invention.

Plato also occasionally uses the Greek terms *on* and *ousia* (both participial forms of the verb 'to be') about items or factors that may or may not be present in the realm of Forms. These words are sometimes translated as 'being' and sometimes as 'existence'; the latter translation seems more appropriate to things in the world of becoming, but the fact that skillful translators (F. M. Cornford, for one) resort to it suggests how difficult it is to deal with what Plato is getting at when he talks of the Forms. Finally, let us return to *eidos* and *idea*, Plato's usual names for a Form: they turn out to go back to the same root, one which yields various words concerned with both seeing and knowing—the same root, incidentally, that produced the Latin *video*. It is plain enough, then, that mind and the Forms are ultimately connected in Plato's philosophy. It is also clear, and not in the least surprising, that the Greeks tend to connect knowing with the seeing of visible forms.

Plato is never dogmatic about the extent of the realm of Forms. In his middle phase—such dialogues as *Phaedo, Republic, Symposium,* and *Phaedrus*—he draws examples of the Forms from morality and art (courage, temperance, justice, wisdom, goodness, beauty); from mathematics (equality, oddness, evenness, triangularity, circularity); from common artifacts (beds, tables, wine jugs, lyres); and from certain qualities, which he sometimes pairs in opposites (hot and cold, sweet and bitter). In a somewhat later dialogue, *Parmenides*, in which he subjects his own doctrine of the Forms to an exhaustive critical examination, he considers the possibility of Forms of biological species and the four elements but leaves that question unsettled. In the same dialogue—perhaps the most difficult of his works—he has the aged Parmenides ask Socrates whether there are Forms of such mundane and even disagreeable things as hair, mud, and dirt. When Socrates rejects that possibility, Parmenides suggests that Socrates is too reluctant to follow out the consequences of the doctrine of Forms (*Parmenides* 170C).

The general point behind all this seems to be that the realm of Forms must be more inclusive than was envisioned in the earlier dialogues, perhaps just as rich as the needs of discourse, of whatever kind, require. It is therefore not surprising that a discussion of divine craftsmanship in a still later dialogue seems to imply that there are Forms of biological species and of the four elements (*Sophist* 265C, 266B). The case for such Forms is even more clear in *Timaeus*, which most specialists take to be a product of Plato's old age. There the Demiurge (the divine craftsman or artificer) is said to construct the world of becoming on the model of an Eternal Living Creature. The setting is mythical or allegorical, but it seems clear enough that Plato is not attributing to that creature both the status of a Form and the characteristics of a concrete living being. He seems to be talking about the *Form* 'Eternal Living Creature', which in context seems to ramify into the Forms that are the originals of all concrete living creatures (30B–D, 37C–E, 39E). In *Timaeus* there is also a discussion of the four elements in which it becomes clear that in constructing the elements the Demiurge must rely on mathematical archetypes that are either Forms or closely related to the Forms (51B–E, 53C–55C). This powerful suggestion, which I consider further in Chapter 3, has been of immense importance in the development of Western science.

Forms as Causes: A Minimal Sense of Causality

For Plato, the Forms are causes of everything that happens in the bodily world of *genesis*, that world of birth, generation, and death we usually call becoming in English. As I remarked in the Introduction, the notions of cause and explanation are in mutual support: if we ask for the cause or causes of something, we are looking for an explanation of it, and that obvious point is probably more salient in the most common Greek word for a cause, *aitia*, than in our own word.

The simplest—and to my mind the least important—sense in which the Forms are causes can be found in one of the earlier dialogues of Plato's maturity, the *Phaedo*. Stated roundly, it is this: describe some concrete and common-sense object in language; if the description is adequate, your language will make reference to a number of Forms in which the object *participates*. These Forms may then be said to be the *causes* of the concrete qualities/attributes of the object described. The Forms are not causes by being immanent in the object, for Plato takes the Forms to be absolutely transcendent in relation to the world of becoming. One might choose to call a concrete quality/attribute an *immanent form*, as R. Hackforth does, but that is not to say, as Hackforth himself makes clear, that the *transcendent* Form is immanent.

The clearest expression of this minimal sense in which the Forms are causes is *Phaedo* 99D–102A, which has been much discussed by Plato specialists. There Plato mentions such Forms as beauty, bigness/greatness, oddness, evenness, duality, and unity; if, for instance, a concrete thing is beautiful, the cause of its concrete beauty is the Form beauty, in which the concrete thing participates. But examples of Forms that Plato gives elsewhere—hot and cold, particular colors—would also fit the present context. Part of the point of this passage is to lay the groundwork for the argument for immortality around which the whole of the *Phaedo* is constructed. The argument, it seems to me, is not a good one, for it requires us to assume that life and death are (transcendent) Forms, and this is hard to reconcile with much of what Plato says about life in later dialogues. The argument runs this way: life and death, regarded as transcendent Forms, are opposites and so exclude each other; so also with the concrete qualities that partake of those Forms; the soul, which brings concrete life to a body, must be deathless, since its coming excludes death. This, however is not the place to criticize the argument: here I am only concerned to draw attention to this minimal sense of causality, which Plato qualifies in many ways in his later dialogues but never repudiates.

Forms and the Purposes of Rational Agency: A Deeper Platonic Sense of Causality

In the middle and late dialogues this minimal sense of the causality of Forms is always accompanied by a sense of telic causality that can only be elucidated by considering further the notion of the Form of the Good on the one hand and the notion of the soul on the other. The relation between these notions brings with it another theme that is of immense importance both for Plato and for the whole philosophical tradition of the West: the causal/explanatory importance of rational agency, whether human or divine. (Recall the brief discussion in the Introduction of the association in Greek thought, long before Plato, between the notions of agency and causality.)

As it happens, one of Plato's most authoritative observations about this complex of notions occurs in *Phaedo* just before the passage I have been discussing. There Plato forcefully insists on the importance and irreducibility of causal explanation that involves rational agency. He maintains that explanation of the human mind in terms of the protoscience of his day is at best a partial explanation, and he makes it clear that he takes the human mind to be an instance of a mind-principle operative in nature at large. But the most significant thing about the passage is its emphasis on the causal authority of the moral component in rational action, an emphasis that clearly associates such decisions with the Form of the Good, although the good qua Form is not so salient in this dialogue as in the somewhat later *Republic*.

The setting of the passage is the Athenian prison in which, on a day in 399 B.C., Socrates was to drink at sunset the hemlock that was to be brought to him by the reluctant executioner. Powerful friends had provided an opportunity for Socrates to escape, and it is thought that the authorities did not wish to make a martyr of him and would have been glad if he had taken advantage of his friends' offer. The story of his refusal of the opportunity on moral grounds is well known; I recount it here only to bring out the doctrine of causality he uses to explain his decision to friends who had gathered to spend the day with him in prison.

At one point in the long conversation, he recalls the scientific and philosophical preoccupations of his youth, and in particular his interest in the teachings of Anaxagoras. He had been attracted to Anaxagoras's doctrine because it purported to be based on the causal power of mind, but on closer examination he was disappointed:

> My wonderful hope was carried away, my friend, for as I went on reading I saw that the man made no use of mind in assigning causes for the ordering of things, but made air and ether and water and other such absurdities causes. It seemed to me much as if someone said that Socrates does by mind all that he does and then, in trying to state the causes of each thing I do, said first, that I am sitting here now because my body is composed of bones and sinews, that the bones are hard and divided by joints, and that the sinews, encasing the bones with the flesh and skin which holds these things together, can be tightened and loosened; and so, the bones being swung in their joints, the tightening and loosening of the sinews enable me to bend my limbs now, and that is the cause of my sitting here bent in this way.

The account so far deals with simple functions and is plainly mechanical; Socrates now extends the imagined account to a more complicated function, and then goes on to introduce a different kind of cause:

> And as if, about the causes of our talking to each other together, he should say voice, and air, and hearing and a thousand such things were causes, and fail to state the true causes, that since it seemed best to the Athenians to condemn me, it has seemed to me best to sit here, and more just to stay and submit to whatever sentence they pronounce. For by the Dog of Egypt, I think these same sinews and bones would have been long ago somewhere in Megara or Boeotia, carried there by their opinion of what is best, if I did not think it more just and seemly to submit to any penalty imposed by the City rather than to flee and run away.

It is absurd, he goes on to say, to call such mechanical things causes of what the imagined opponent has characterized as an act of the mind. Socrates concedes that he could not have carried out his decision without bones, sinews, and all the rest: the mind in question is, after all, an embodied one. But he insists that whoever supposes such a causal account to be an adequate one has missed a vital point: "the real cause is one thing, and the cause without which the (real) cause cannot be the cause is another" (*Phaedo* 98B–99B).

This is the first clear occurrence in philosophy of the distinction between a cause that is sufficient to produce its effect and an accessory, auxiliary, or conditioning cause—one that is merely necessary for the sufficient cause to do its work. The word 'necessary' is often invoked in discussions of causality, and as it has more than one shade of meaning, it is a source of some confusion. One possible confusion is that once all the conditions for the operation of the sufficient cause are in place, the sufficient cause is sometimes said to *necessitate* its effect—that is, to produce its effect inexorably. The doctrine of determinism, which I discuss in Chapters 3 and 4, invokes that sense of necessity. Plato has in mind a conditional necessity, one that subserves a purpose of mind. Purposes of mind, in his view, are always dependent on the Form of the Good and so do not operate with necessity. (The Form of the Good is not explicitly introduced in the passage I have just quoted, and indeed it is explicitly mentioned only once in the *Phaedo* [100B]; but just after the passage quoted Socrates remarks that those who embrace the mechanical version of causality have ignored a *power* that places things as it is *best* for them to be placed.) Socrates thought that he *ought* to stay in prison, but this 'ought' carried with it the possibility that he might have failed to accept the persuasion of the Good: it was open to him to choose flight instead. On the other hand, the decision once taken, we might then say that the *decision* necessitated—that is, fully accounted for, was sufficient for—his staying and his acceptance of the hemlock.

The Form of the Good and the Causal Role of Soul

Plato thinks of the realm of Forms as an ordered or structured one: the human mind can explore it with the help of the conversational method he calls dialectic, and the result is systematic knowledge expressed in logical discourse. This is a correlate of the notion that the Forms are above all objects of knowledge—that they are *what* we know *when* we know. But from the *Phaedo* passage we have just looked at, it is clear that there is a certain ambiguity introduced into this ideal of linguistic clarity and expressibility by the dominant presence of the Form of the Good among the other Forms. Its presence makes the Forms a hierarchically ordered realm in which the order is also a moral one rather than merely one of logical coherence and degree of generality. This moral order has, moreover, a telic causal significance, for the Form of the Good is the eternal correlate within the Forms of a self-moving principle called the soul. In such mature middle-period dialogues as *Republic*, *Symposium*, and *Phaedrus*, which come just after *Phaedo*, the soul is the life-principle of all individual organic things; and though it remains so throughout Plato's work, the notion of individual souls is supplemented in certain later dialogues by the notion of a world-soul.

The appearance of the world-soul is foreshadowed by a passage in which Plato attributes the origin of all animals and plants to divine craftsmanship (*Sophist* 265B–C). The world-soul makes its explicit appearance in a remarkable but difficult passage in which Plato sets a constructive divine mind within a world-soul (*Philebus* 28D–30E). The story culminates in the systematic cosmology of *Timaeus*, in which a world-soul constructed by the Demiurge is made the most important ordering principle of the world—a telic order whose presence is felt even in what we should call today the ordinary physical and chemical processes of the world (*Timaeus* 34A–40D). The agreement of the discussion of conditioning or auxiliary causes in this late passage with Plato's early discussion of conditioning causes in *Phaedo* is remarkable.

Everything that Plato says about the Form of the Good suggests that although it is the source of all intelligibility and definability, it is not itself intelligible and definable in the same sense. He tells us that the other Forms derive their being and intelligibility—that is, what makes them objects of knowledge—from the Form of the Good, but that the Good itself is not an object of discursive knowledge. The Good, he says, is beyond truth and knowledge and indeed beyond Being itself (*Republic* 508E–509B). It is therefore natural enough that somewhat later in the same dialogue Socrates begs off from giving a complete account of the Good. For the same reason, the two famous accounts of the upward progress of the soul in *Republic*—the schematic one of the Divided Line and the concrete and poetic one of the Allegory of the Cave—do not culminate in the discursive function of the mind (*dianoia*), but in the directness of rational intuition (*noēsis* or *nous*), on which the discursive function is said to depend.

The telic significance of the Form of the Good is the obverse of its status as a final explanatory principle that cannot itself be explained. That is why the Good is at least implicitly present when Plato talks of the gods and especially when he introduces the notion of a divine constructor. In *Timaeus*, as noted above, a god-like figure shapes the visible world by taking the Eternal Living Creature as his model. Although the Form of the Good as such is not mentioned there, goodness is still the controlling factor, for that model is chosen by the Demiurge because it is the best model. Myths aside, the Form of the Good appears to be an impersonal deity principle whose ambiguous and puzzling presence among the Forms makes the realm of Forms an ideal that the world of becoming strives towards but never quite achieves.

The Development of Plato's Views on Soul

We are now ready to look in more detail at the role of soul in Plato's philosophy. Plato's views are complex, and they also develop in the course of his long philosophical life. At one extreme the soul seems to be identified with mind, so that its true home seems to be in the realm of Being and its true task to know the Forms and whatever else is resident in that realm. That is a theoretic task in the literal sense of the verb *theōrein*, which means 'to look on', as a spectator in the Olympic games looked on but did not participate in the games. This way of regarding the soul, which Plato may well have taken over from the Pythagoreans, makes any concern of the soul with the region of becoming an alienation from its true task. When the soul is embodied, it is imprisoned, and much of the account of it in *Phaedo* is consistent with this interpretation. Even the sensory apparatus, which in that dialogue is given the positive function of reminding the soul of what it already knows from its true home (the famous doctrine of *anamnēsis*, or remembrance) is a temporary dodge, needful only for the soul's imprisoned state. In its true home, confronted with its true objects, it simply contemplates the Forms without the aid of sensation but also without the disadvantage of that distorting lens.

In the rather more detailed discussion of the upward progress of the soul to knowledge in *Republic*, the discursive and argumentative function of reason (*dianoia*)—a function that frequently has recourse to diagrams and other examples tied to the realm of sensation—gradually gives way to a more direct function for which Plato uses such words as *noēsis* and *nous*. These words are translated in various ways; one of them, 'rational intuition', reminds us that the function is closely related to the one Kant calls intellectual intuition. It is a central tenet of Kant's doctrine that human beings do not possess such a function.

Even as early as the *Phaedo*, in which the language is dualistic and the soul seems alienated from its proper state when it is in the body, there is another side to the doctrine of the soul. Whatever we may think of Plato's argument for the immortality of the soul there, it is clear that he supposes that the soul brings life to the body. But it does more than that, for part of its mission, according to Plato, is to rule or master the body and by doing so produce a complex of soul and body that is virtuous—in short, a good person (79E–80A). If soul were indeed in its deepest nature merely an imprisoned theoretic mind, it could hardly do that. We are told in *Phaedo* of the binding force of the Good, which holds all things together (99C). The soul, in Plato's view, has something in it that can make common cause with this very concrete aspect of the Good, if only, as Plato says, by opposing the body—sometimes forcefully and even harshly, sometimes more persuasively (94B–D). An alliance with the concrete power of the Good, together with so concrete a mastering of the body, seems to require a concrete footing in the body.

In the *Republic*, as everybody has read at one time or another, this concrete footing is made explicit in the doctrine of the three parts of the soul—appetitive, spirited, and rational, with the rational part being subdivided in accordance with the distinction between discursive reason and rational intuition (434D–441C). The parts are intimately related: though appetite is assigned to the lowest part of the triad, the other parts have inclinations or desires that echo the theme of appetite. The rational part desires to know the Forms, and especially the Form of the Good. The spirited part desires to make common cause with the rational part, but it can be so perverted that it turns its dynamism towards the gratification of the appetitive part. The doctrine of the three parts of the soul is restated in *Phaedrus*, with considerable poetic license, in the myth of the winged charioteer who drives two winged horses, a white and generally orderly one and a fractious dark one. The charioteer represents the rational part, the white horse the spirited part, the dark one the appetitive part (246A–247C).

In *Symposium* and *Phaedrus* all the earlier material about the telic relation between the soul and the Good is reshaped in terms of Love, and the Form of Beauty becomes the vicar of the Form of the Good. The telic attractiveness of the realm of the Forms manifests itself first in bodily beauty, and when the ultimate source of that beauty is understood, the soul ideally goes on to make common cause with the being of the Forms, of which bodily beauty is at best only a reminder. Often enough the soul does nothing of the kind. The images in *Symposium* and *Phaedrus* are concrete enough, excessively so for some tastes. Those dialogues—especially *Phaedrus*—remind us that for Plato madness is just as important as method. It is madness, to be sure, ideally defined as divine madness—*mania apo theōn* (madness from the gods) or *enthousiasmos* (having a god within); but it is madness, right enough, or at least inspiration, and its dynamism seems to carry us towards art rather than towards philosophy. The tension in Plato's own nature between art and philosophy is plain enough in these dialogues, despite the many negative things he says about art in the *Republic* and elsewhere.

Phaedrus, however, is also important for another reason: it gives some precision to the notion that the soul brings life to the body. Plato tells us now that the soul can do so because it is self-originating movement—movement in a sense broad enough to include the complex movements at the microlevel we call metabolism. All living things possess this character of self-movement; all nonliving things lack it, and what motion they exhibit has its origin in soul (245C–246A). Plato is thus able to provide here a more plausible proof of the immortality of the soul than the one he gives us in *Phaedo.* For one thing, the soul does not merely rule the body in the *Phaedrus:* it now takes care of the body, and not merely the living body, for its care now extends to all inanimate bodies as well.

Moreover, a theme that was only implicit in the reincarnation doctrine of the *Phaedo* becomes explicit in the *Phaedrus:* Plato universalizes the notion of soul in the passage just cited. He now uses the phrase *all soul:* "All soul is immortal" (*psychē pasa athanatos*), he says (245C); and a little later, "All soul takes care of that which is without soul" (*pasa hē psychē pantos epimeleitai tou apsychou*) (246B). Though the myth in which these observations are set also speaks of multitudes of mortal and immortal souls, the notion of *all soul* seems dominant. It is said to traverse the whole of heaven, "now in one form, now in another," an image which offers small comfort to those who want assurance of personal immortality.

Though soul in the *Phaedrus* does not have its immortal status in the region of becoming but rather in the region of Being—which comprises Forms, the divine, and (in still later dialogues) divine agency—soul is clearly the vicar of Being in the region of becoming, since it is said to take care of everything that resides there. Accordingly, the language Plato uses about becoming takes on a more positive tone from that dialogue on. Becoming is no longer simply the region in which things are generated and pass away: it is now the region in which the telic power of soul works, and soul brings to it something of Being.

This is very clear in Plato's story in *Timaeus* of how the Demiurge makes the world-soul and then makes the souls of living things out of a diluted version of the ingredients of the world-soul (35A–B). The three basic ingredients—being, sameness, and difference (otherness)—are called intermediate: intermediate being, intermediate sameness, and intermediate otherness/difference. The intermediate nature of these Forms consists in their having a portion (here called indivisible) that belongs entirely to Being and a portion (divisible) that belongs to becoming. These two portions parallel, but only roughly, the earlier distinction between a Form as absolutely transcendent and an immanent "form" (properly, a quality or predicate) that bears the name of the Form.

The rest of that complex myth I cannot go into here. But I must at least make the important point that the world-soul is constructed out of a mix of ingredients which includes at least three components that come straight from the realm of Being: a pure portion of the Form being, a pure portion of the Form sameness, and a pure portion of the Form otherness/difference. A more positive note about the importance of becoming could scarcely be sounded without departing entirely from the Platonic vocabulary.[1] Just as positive is the famous passage in *Timaeus* in which Plato calls time a moving image of eternity (37C–38C). This positive note carries over even to what was earlier called not-being. Now, considered as undifferentiated space, with which the Demiurge works and in which he shapes the world-soul and individual souls, it is given new names—the Receptacle; the nurse, or mother, of becoming—and so is no longer regarded as utter not-being but only as almost not-being. It thus *partakes* of the intelligible (*tou noētou*) (51A), which is to say that it has some relation, however difficult and tenuous, with the Forms; indeed, though the Receptacle has no fixed character of its own, it is said to be apprehended by a kind of bastard reasoning that does not involve sensation (48E–49A, 50D–52C).

Plato is a many-sided philosopher, and there are other factors besides the development of his doctrine of soul in this tendency to use the term 'being' in some qualified way about becoming as he grows older. One such factor is his recognition, in *Parmenides, Sophist,* and *Timaeus* itself, of a *Form* 'being' within the region or level of Being and his systematic examination of the importance of that Form for all rational discourse—discourse not just about the realm of Forms but about becoming and not-being as well. In *Sophist,* for instance, it is conceded that even an image, which belongs at the level of becoming, has being "somehow" (*on pōs*) precisely because it *is* an image (240B).

Plato's Views on the Causality of Mind

In Plato's version of the ancient doctrine of hierarchical causality, form and purpose (telos) are central—central even to the very existence of becoming—as this brief examination of the notion of soul should make clear. But we must not forget that rational agency itself, from the famous account of Socrates' human agency in *Phaedo* to the account of the works of the Demiurge in *Timaeus*, is also central. This means that for Plato mind, or reason, is not something for which we need to give a causal explanation: it is itself a magisterial cause that must be invoked in any adequate explanation of other things.

It is clear that even as early as *Phaedo* Plato thinks that mind, or reason, is the highest functional level of the soul. But as we trace the development of the notion of soul through such later dialogues as *Sophist, Philebus,* and *Timaeus*, we find that a *divine* Mind is not just the highest level of the world-soul (or, if you prefer, soul regarded as a universal principle) but also the maker of that soul and so the maker of individual souls as well.

The passages in question are difficult enough, but they are also decisive. In *Sophist* all living things—besouled things—are attributed to divine craftsmanship working with reason and knowledge (265B–C). In *Philebus* Plato uses a fourfold scheme in a fresh approach to the contents of his usual threefold scheme of Being, becoming, and not-being. The new scheme is made up of the Pythagorean notions of the *limit* and the *unlimited*, supplemented by the *mixture* of those two and the *cause* of such a mixture. In an obvious sense, the limit is the realm of Forms, and the unlimited is not-being. But Plato also applies these notions to the Forms themselves. A given Form (say, 'animal') may be both a limit and an unlimited: as *one* Form it defines (i.e., limits) the class of beings that participate in it; as awaiting further limitation by the Forms of the various kinds of animals, it is a potential *many* and so is indeterminate or unlimited.

Just here, however, I set aside that aspect of the scheme to draw your attention to the application of the scheme to the relation between Being and becoming. As the region of soul, both the world-soul and particular souls, becoming is a mixture of the limit and the unlimited. The cause of that mixture, however, is said to be a divine Mind (*nous*). It appears, then, that mind, which from one point of view is the highest functional level of soul, from another point of view is a transcendent or universal Mind that makes the soul in which it is also immanent (*Philebus* 15D–17A, 23C–30E). Much the same argument can be made about *Timaeus*, in which, as I noted earlier, there is an extensive account of the making of the world-soul and all individual souls by a divine craftsman who works from a model he knows to be best.

It is hard to overstate the importance for later philosophy and religion of this Platonic account of a divine Mind that shapes, in accordance with a supreme ideal of goodness and a subordinate ideal of beauty, the nature and course of the universe. Though in Plato's story that mind is confronted with factors it has not created, it orders the universe purposefully and builds into it subordinate telic orders of a most ingenious kind. Plato's account is the ultimate source of the West's vision of God as a divine artificer or constructor, right down to the homely image of a workman: the Greek word *dēmiourgos* means no more than 'worker for the people', although Plato's myth has endowed the word with an air of mysterious power that makes us tend to capitalize it. The transformation of Plato's great invention by later theological developments is described briefly in Chapter 3.

Aristotle's Revision of the Platonic Doctrine of Forms: Form as Essence

For Aristotle, as for his master Plato, philosophy is properly concerned with the study of being, and in that study the topic of form (*eidos*) is of the first importance. But there are notable differences between the two philosophers. Aristotle's primary example of being is not a form considered in itself, but rather a particular, individual, changing, and developing being (*ousia*)—a man, for instance. It is true that if you ask what makes that man real or actual—what, in effect, makes that man a being—Aristotle's answer is that it is the form that makes him so. When he discusses form understood in this sense, Aristotle often supplements the Greek *eidos* with certain Greek expressions that emphasize the notion of being. These expressions, when translated by way of the corresponding Latin expressions, bring into philosophy the important notion of *essence*; I consider these expressions somewhat later.

Let us now call the form that makes something a real, or actual, being an *immanent* form/essence. To do so is to suggest that Aristotle's doctrine contains a puzzle that is the obverse of Plato's: for Plato there is a puzzle about how transcendent Forms can be participated in by things in becoming; for Aristotle there is a puzzle about how a form that seems to be immanent in an individual—so much so that the very actuality of the individual depends on it—seems also to transcend that individual concrete setting. For Aristotle, if form makes Socrates, Plato, and Aristotle individual concrete beings, it is not three distinct forms—a form of Socrates, a form of Plato, and a form of Aristotle—that do the job. It is the form *man* (or, if you prefer, the form *rational animal*) that does it.

But if the word 'immanent' does not cover every aspect of Aristotle's forms/essences, whatever word we do hit upon must take into account the very concrete operation of form, and it must also take into account the fact that form is less than fully real—less than *actual*, to use the word that commonly translates two key Aristotelian expressions—when it is not operating in that concrete way.

Casting about for a more dynamic term than 'immanent' and one that does not immediately entangle us with the transcendent-immanent distinction, I recalled some lines from Dryden's *Absalom and Achitophel*, lines which also remind us that for Aristotle the form of a living body is the soul of that body. Dryden, as it happens, is writing about a person he takes to be disreputable, but that does not detract from the appositeness of the expression he uses:

> A fiery soul, which working out its way,
> Fretted the Pigmy-Body to decay;
> And o'r informed the Tenement of clay.

An informing form, or essence, has some of the marks of a Platonic Form: it can be defined (Aristotle's usual definition in the case of man is 'rational animal'); it is not particular (it is not the form 'Socrates' or 'Plato' or 'Aristotle'); and it is an ideal (any of those men can fall short of it as they live their lives). Finally, it has a causal significance, for it is the reason why the individual being develops into a man by a long process of change—change that does not affect the form itself, which, like a Platonic Form, merely *is*, at least when it is doing that concrete informing job.

The same form, however, can be abstracted by a knower and thought about, in which case it is no longer the causally vital form (the informing form) of some particular being but rather a universal form (the form 'man' or 'rational animal'). We are tempted to say, as we are not in the case of a Platonic Form, that in such cases it is a *merely* universal form. At any rate, it is, qua known, not fully actual. All this being said, I shall now feel free to use 'informing form' and 'immanent form' interchangeably in contrast with 'abstract form' or 'universal form'.

Aristotle's usual word for beings like Socrates, Plato, and Aristotle is *ousia*; it is a (feminine) participial form of the verb 'to be', and the simplest way to translate it, as translators are only now beginning to realize, is the obvious one: 'a being'. Note the presence of the article: 'a being', rather than simply 'being'. Aristotle is not talking about being but about many individual beings. When he wants to talk about being—the being that all beings share—he uses another (neuter) participial form of the verb 'to be' (*on*), which is the same word Plato uses about the Forms. Most of the individual beings Aristotle considers are developing, changing beings, and it is clear that he is most at home with biological beings, but in principle the term *ousia* can be applied to static beings as well.

With the important word *ousia* before us, we are in a position to be more precise about the difference, in Aristotle's view, between an informing form and an abstract form. In an early Aristotelian book, the *Categories*—which some Greek scholars now think was not actually written by Aristotle, although it is clearly consistent with works about which there is no such doubt—an individual man is called a primary *ousia* and the abstract (universal) form 'man' is called a secondary *ousia* (*Categories* 2a11–3b24). Form as abstract/universal does not do or accomplish anything, and so it is secondary.

This changes nothing that was said just above about the importance of form as immanent: an immanent form *does* something. If the individual is a primary *ousia*, it is the form immanent in the primary *ousia* that makes it a real, primary being, so much so that Aristotle calls this informing form the actual *ousia*. The Greek expressions he uses to make his point are *ousia energeia* (*Metaphysics* 1042b11, 1072a25–26) and *ousia entelecheia* (*On the Soul* 412a20–23, 412b5–6). Though both *energeia* and *entelecheia* are usually translated as 'actual', the first term obviously connotes activity, the second completeness. Remember, however, that the active, actual, informing principle which makes concrete things real is, so to speak, itself fully real (actual) only when it is doing that job.[2]

One final terminological point will be helpful before we turn to Aristotle's views on soul and mind. As I said at the beginning of this discussion of Aristotle, immanent, or informing, form—the very actuality of the primary *ousia*—is often referred to in Aristotle's writings by two Greek expressions that are usually translated as 'essence'. They are *ti esti* and *to ti ēn einai*; they mean literally 'what is it' and 'the what it was to be' respectively—the latter as odd in non-philosophical Greek as it is in English. These odd expressions are, as it were, saturated with the verb 'to be' (as, of course, is the term 'essence'): they draw attention to what Plato called Being, but the way Aristotle uses them to discuss what he takes to be immanent in particular beings brings out the distinctive Aristotelian note.[3]

For Aristotle particular changing things are *beings* (*ousiai*). Aristotle's modified Platonism consists in this: the very actuality of a being is *form;* the active informing principle which makes it a concrete and functioning being worthy of the expression *ousia energeia* is also that which makes the being intelligible and thus capable of being defined. In short, the form that informs a primary *ousia* and indeed makes it actual and thus primary also endows it with intelligibility and so makes it capable of providing to the inquiring reason that secondary *ousia* we call a universal or abstract form. But in the end it is a modified Platonism, for a Form is anything but secondary in Plato's philosophy.

Aristotle on the Soul and Its Causality

Aristotle's clearest and most detailed example of the actual, informing form/essence of a primary *ousia* is the soul. As the actual form of a body that is potentially living, it informs such concrete individuals as human beings, and so brings *being* to them, and the being it brings is life. Though the soul is a life-bringer, it is nonetheless a *form* (though not precisely a Platonic one), and so is also the principle that makes the human being intelligible. If we wish to define or otherwise consider rationally this vital source of the being of a primary *ousia*, and if we wish to take account of the fact that although it is the vital source of individuality, it is not itself an individual, we must perform an act of abstraction. Once we do this we are, in the language of the *Categories*, considering a secondary *ousia* rather than a primary *ousia*: we are considering rational animality in general.

But the Greek expressions that produce the translation 'essence' are sometimes applied both to the informing form in its concrete actuality and to the same form as entertained and defined by our intelligence (the abstracted form), and this is a source of some confusion. It helps to remember that when Aristotle wishes to single out the abstract form, he often calls it a universal: so, for instance, at *Metaphysics* 1038b10–19, which I cited above. Soul does work; it is a cause; and though in the case of a human being the structure of that soul makes it appropriate to define a human being as a rational animal when we are asked what the essence of a human being is, we are nevertheless unlikely to confuse that essence qua definition with the essence that gives the human being the movement and activity we call life.[4]

Master and pupil though they are, Plato and Aristotle differ in certain important respects about the nature of the soul. For Plato, the soul is a self-mover, though not an ultimate one, for it is moved by desire, or love, of the Forms and in particular the Form of the Good; and as a self-mover it brings life to the body (or the many bodies) with which it is joined for a time. For Aristotle, the soul is the principle of movement for the body's growth and development as well as its mature actions, but it does not itself move, except incidentally, as the man of which it is the actuality moves himself about. Although Aristotle did not believe in a realm of independently existing Forms, he took the soul to be the informing form of the body—in effect an unmoved mover (though not the ultimate Unmoved Mover) of the body, providing the source of, and the goal for, the body's movement and development.

As such, the soul is a cause of a most important kind. Aristotle calls the soul the cause of the body in three of his four senses: formal, telic/final, and efficient (*On the Soul* 415b9–29). The formal cause we have already considered at some length, and it should be clear that the soul, considered as formal cause, is complete (actual) and unchanging. But that same informing form does its work only by being the goal or end (hence, telic or final cause) for the process which is the development and the completed functioning of the rational animal. The soul is therefore the cause of all the movement—both the movement that is development and the movement that is locomotion—of the rational animal (hence, moving, or efficient cause). The soul is not, however, the source of the material (potentiality) which it informs (actualizes): it is not, that is to say, a cause in Aristotle's fourth sense, the material cause. Nor is it the ultimate source of the formal, final, and efficient causality it possesses, for there is a God in Aristotle's universe, although a God whose role is significantly different from the one played by Plato's God.

Aristotle on Rational Agency

There is in Aristotle a complete and subtle account of human agency. It is found chiefly in the *Nicomachean Ethics*, which is as necessary to an understanding of his doctrine of the soul as is his treatise *On the Soul*. Aristotle takes agency to be causally effective both in the world and in the proper development of the soul itself, for it is virtuous conduct that develops the soul towards its ideal condition. A paradox, that, because as we have just seen the soul is the cause of the living body (hence of the living person as well) in three of the four senses of 'cause'. We can partly resolve the paradox by noting that the soul has three parts or levels: a rational part, a sensitive (or appetitive) part, and a nutritive (or vegetative) part. The first two correspond roughly to the three parts of the Platonic soul, for what Aristotle calls choice (*proairesis*) is the key to virtuous conduct and hence to the proper development of the human soul, and choice has something in it of both upper parts of the soul.

Aristotle calls choice reason that desires or desire that reasons, and it is clear that, like the Platonic spirited part of the soul, it is both perfectible and corruptible—can ally itself, that is, either with reason or with appetite/desire. Moreover, the soul, regarded as a (universal) form that is the basis of definition no less than of life, supplies the definition 'rational animal' and so contains within it a tension between two components. The third component of the soul is implicit in this tension, for animality is actual only when metabolism is going on, metabolism being a fair modern equivalent for what Aristotle calls the nutritive/vegetative soul.

We can complete the picture by noticing that, for Aristotle, form and matter are relative notions. If, for instance, the animal level of soul is the form of the material that is the nutritive level of soul, the animal soul is on the other hand the material of which, in the case of human beings, the rational level of soul is the form. My resolution of the paradox of the development of a soul that is the causal source of biological and moral development is perhaps trivial: I have merely pointed out the tension that is the source of the paradox. One way of expressing the tension is to say that soul is the ideal that informs all living actualities whose actuality is an actualizing—that is, a becoming.

The diagram shown here expresses the main argument of the *Nicomachean Ethics* in terms of the structure of the soul, its parts, and its virtues (*aretai*). Notice that the parts can also be construed as functional levels: each part has its particular work/function (*ergon*) to perform, and the virtue of any particular function is simply the excellence (*aretē*) of that function. If we try to give a formula for the virtue in general of a rational animal it is simply *living well*, but as a rational animal is a creature made up of several functional levels (since rationality itself is subdivided), living well is a complex notion. Ideally it culminates, as it did for Plato, in philosophical thought, but the virtues of rational *animality*—a virtue like courage, for instance—are necessary, though not sufficient, for the achievement of the philosopher's wisdom. Courage and wisdom are thus related as matter and form respectively.

Although for Aristotle the human agent works towards a God-like life of pure thought, it does so only by way of prior actualizings of embodied levels of its

The Structure of the Soul in Aristotle's Ethics

Parts/functions of the soul	Sub-parts/functions	Virtues
		INTELLECTUAL VIRTUES [*dianoiētikai aretai*]
Rational [*logon echon*] 1102a29–30	Theoretic/comtemplative/scientific [*hē theōrētikē dianoia*]1139a28 [*to epistēmonikon*]1139a13	Rational intuition [*nous; dianoia*] Scientific understanding/ demonstrative knowledge [*epistēmē*] } Wisdom [*sophia*]
	Calculative/ deliberative [*to logistikon*] 1139a13 { Practical [*praktikos*] Productive [*poiētikos*] → CHOICE [*proairesis*]	Practical wisdom/prudence [*phronēsis*] (true reasoning with respect to making)
Sensitive/ appetitive Nonrational [*alogos*] but "somehow participating in reason" (1102b)	Desire → Appetites Sensation Locomotion	**MORAL VIRTUES** [*ēthikai aretai*] Courage Temperance Liberality Friendliness etc. "Moral virtue is a habitual disposition towards choosing that which is moderate with respect to our (particular) selves—moderate as defined by such a principle as a man of practical wisdom would lay down. (1106b36–1107a2)
Nutritive/ vegetative Nonrational	Nutrition Growth etc.	Health

being. At those levels practice—in the root sense of doing things in a physical world—dominates, although standards accessible to mind must be invoked if the actualizations are to involve the virtues/excellences proper to those levels. To put the matter in a more down-to-earth way, it is particular choices directed towards the ideal of moderation in the realm of the appetites/desires that furnish the material conditions for the eventual actualization of a contemplative rational life. As noted above, choice is reason that *desires*, desire that *reasons*.

Aristotle's God is indeed the telos towards which the life of the ideal moral agent moves, but in Aristotle's theology we find no clear analogue of the divine craftsman (the Demiurge) that is so central to the later work of Plato. Aristotle's God is not an agent, at any rate not in that sense of 'agent'. The logic of Aristotle's thought requires that his cosmic hierarchy of levels in which form and matter are relative notions should be closed at each end by absolutes—by prime matter at the lower end and the Unmoved Mover/Prime Mover at the upper end. The logic requires also that prime matter should be pure potentiality and the Prime Mover pure actuality, because for Aristotle matter is potentiality and form is actuality. Aristotle does indeed attribute life and mind to God, but he tells us that God lives a life that is pure thought—thought that thinks about thinking: "mind must think itself if it is that which is most good; it is a thinking that is a thinking of thinking" (*auton ara noei, eiper esti to kratiston, kai estin hē noēsis noēseōs noēsis*) (*Metaphysics* 1074b34–35). In a telling related passage Aristotle remarks that the actuality of thinking is life and that thinking is God's actuality (1072b27–30). Since what makes that life pure actuality is also what makes it exempt from all actualizing, the life God leads cannot be life in the sense of human life.

When Aristotle calls living beings primary beings (*ousiai*), he must therefore mean that their primary status is also relative, for no such beings can carry out their actualizings on their own. They are dependent on the pure actuality of the Unmoved Mover, which is the ultimate source of the movement that is essential to their actualizing. But the Unmoved Mover is an efficient cause only in the sense that its self-contained thinking life is the telos at which all movement aims. There can be no first efficient cause in the sense of a first movement for Aristotle, because he argues, in certain famous passages in the *Physics* and *Metaphysics*, that physical movement and time have no absolute beginning or end. So God is the first efficient cause only by being the first telic cause. Is God also the first formal cause? Clearly he is not, if to be so is to actively create the forms of finite changing things. Still, all actualizings would fail, according to Aristotle, if there were not a being that is eternally actual; and form *is* actuality; so, in some sense Aristotle does not make clear and does not directly address, it would appear that for him God is also the first formal cause.

If we seem remote at this point from Plato's Demiurge, we are not at all remote from Plato's Form of the Good, for Aristotle's account of God's causality depends on the notion that God's goodness operates as a telic/final cause—the proper object of both thought and desire. God, so understood, causes all the movement in the world by being loved, not by moving and so transmitting motion (*Metaphysics* 1072a26–b31).

There is at least one lesson we may draw from all this: the ancient hierarchical view of causality of which Plato and Aristotle are the most influential representatives makes mind an explanatory principle of the first importance for the universe at large. Although both philosophers acknowledge that the doings of finite minds are in part dependent on lower functional levels of the soul, and so on what I called an infrastructure in the Introduction, they both insist that explanation in such terms is only partial and that finite minds share in the irreducibility both philosophers take for granted in the case of a Mind that is not finite and particular.

2

Descartes's Dualism and Its Disastrous Consequences

The Extreme Nature of Descartes's Dualism

DESCARTES'S DUALISM is an extreme one: as he understands the two things for which the term 'dualism' was invented, mind (or soul) is radically different from body (or matter), so different indeed that it seems preposterous to think of the two things as joined. Each of the two seems capable of an existence on its own; each seems to have no need of the other. Moreover, each seems so constituted as to preclude any causal effect of one on the other. Yet Descartes argues that in the unique case of man these two things are conjoined and indeed interact. Descartes is in some debt to Plato by way of the Christian Platonism of St. Augustine, but the dualism of Plato, which is more definite at the time of the *Phaedo* than in the later dialogues, was never so extreme as that of Descartes.

The most detailed and impressive exposition of Descartes's dualism is to be found in his *Meditations on First Philosophy* (1641/1990). It is a complex book, and I do not intend to paraphrase its argument in detail but only to pick out the themes and arguments that define Descartes's dualism and contribute to its later influence. Innovative though the book is, many of its arguments are expressed in a vocabulary inherited from medieval Aristotelianism, so it is important to notice at once two respects in which Descartes differs from Aristotle.

Descartes's Rejection of Aristotle's Doctrine of the Soul

The first difference between Descartes and Aristotle concerns the soul. Descartes holds that the lower two parts (or functions) of the tripartite Aristotelian soul are not necessary to explain the functioning of a living body, whether the living body is that of a man or that of an animal—a brute, in Descartes's vocabulary. The three parts/functions of the soul are often simply called souls, and in most translations the two lower parts appear as the sensitive/appetitive soul and the vegetative/nutritive soul. Aristotle believed these two lower souls were indeed necessary, because for him the soul-principle was precisely the life-principle; in his view, man has a sensitive/appetitive soul in common with the other animals, and all animals have a vegetative/nutritive soul in common with plants, but the rational soul is unique to man.

Descartes was fascinated from the very beginning of his career by the notion that a living body can profitably be understood as a machine, or automaton, although one devised by God, the Perfect Being. In an early work from the period 1629–33, the "Treatise on Man," Descartes denies the existence of the two lower parts of the soul. He first discusses the many machine-like functions of the human body and then goes on to say:

> In order to explain these functions, then, it is not necessary to conceive of this machine as having any vegetative or sensitive soul or other principle of movement and life, apart from its blood and its spirits, which are agitated by the heat of the fire burning continuously in its heart—a fire which has the same nature as all the fires that occur in inanimate bodies. (Descartes 1991, 1:108)

Although Descartes did not publish the "Treatise on Man," this thesis became a permanent part of his philosophy. He mentions it again in 1637 in his *Discourse on Method,* part V, in which the machine image is once more applied to man. After comparing the human body with man-made automata, he remarks that the body, regarded as a machine made by God's hand, is incomparably better organized and has more admirable movements than things invented by man (ibid., 139). The same doctrine persists in part I, articles 5 and 6 of his *Passions of the Soul,* which was written some four years before his death (ibid., 329–30).

From all this we may conclude that Descartes dismisses the two lower of the three Aristotelian souls and that he identifies the human soul with the mind, that is, with the Aristotelian rational soul. This means that for him any interaction between mind and body must be understood to take place between a mind/soul on the one hand and a machine-like physical body on the other hand.

A Profound Difference between Descartes's Doctrine of Substance and Aristotle's Doctrine of Ousia

The second difference between Descartes and Aristotle concerns the meaning of the term 'substance'. Descartes's claim that mind and body are radically distinct in nature is presented in the *Meditations* as a claim that they are two distinct substances that could, in principle, exist in independence of each other but in the case of man are combined in one nature—a composite nature, as he calls it in Meditation VI. The difference between Aristotle and Descartes on this issue is not a mere matter of terminology, but there is a terminological point we must deal with first if we are to understand the real issue that lies behind it.

The word 'substance' (*substantia* in Latin) has been the standard translation of Aristotle's *ousia* (a being) since at least the time of the acceptance of the Athanasian creed by Western Christianity in the course of the fourth century. The reasons for this need not distract us from the point at hand: what is important here is that it is a poor translation, poor because it seizes upon just one aspect of an Aristotelian *ousia*, namely, that it persists as a continuously existing entity/being through whatever changes it undergoes. But in Descartes's hands, this identification of *ousia* with just one of its aspects is complicated by his failure to notice that an Aristotelian *ousia* is an *individual* being/entity. If we follow Descartes in this use of the term 'substance' we can easily lose sight of the important point that a human being is Aristotle's favorite example of an *ousia* and so (given the usual but misleading translation of *ousia* by 'substance') should be a legitimate example of substance. Descartes, while retaining the usual Latin equivalents for the Aristotelian vocabulary, is in fact talking about something else: a Cartesian substance is not an Aristotelian *ousia*.

Descartes takes mind and extension as his examples of finite substances. His notion of extension is that of a three-dimensional continuum such as a mathematician might define, analyze, and regard as the locus in which mathematical figures might be constructed. But extension in that sense is precisely what Descartes means by body or material. His views on material thus remind us of the Platonic doctrine of the Receptacle, which is discussed in Chapter 1; but I am not aware that Descartes mentions the Receptacle anywhere in his works.

It is important to notice that it is extension in general that Descartes regards as material substance: he does not mean that some particular bit or region of extension is substance, nor does he mean that some particular figure constructed in extension is substance. In the synopsis that precedes the *Meditations* he makes that point explicitly and then goes on to say that any particular human body is just a temporarily persisting configuration (a particular set of "accidents") of extension (Descartes 1991, 2:10). In the same context he points out that mind/soul is a pure substance rather than a particular configuration of accidents of a more general mind substance; thus a mind does not become a different mind by changing its ideas, sensations, wants, and so on. Later philosophers were less willing than Descartes to exempt the mind from the generalizing tendency of his notion of substance.

The Thinking Thing Alone with Its Ideas

In Meditations I and II Descartes establishes his well-known first certainty: the existence of himself, understood as what he calls mind, soul, thinking thing, or thinking substance. One tends to think of the argument that establishes that certainty as the Cogito and to sum it up in the formula *Cogito ergo sum.* That formula, however, appears in the *Discourse on Method,* not here in the *Meditations on First Philosophy;* here Descartes simply says that the statements *Ego sum, ego existo* are necessarily true every time he pronounces them or conceives of them. To pronounce them is to think; to conceive of them is to think; and thinking is the thinking being's way of existing. Here thinking is to be understood in the broad sense of the well-known list in which Descartes defines a thinking thing as one that doubts, understands, affirms, denies, wills, is unwilling, imagines, and senses. (I suggest a more extensive list in Chapter 5.) Doubt and error cannot undermine the initial certainty; indeed they reinforce it in the spirit of the formula of St. Augustine, the philosopher who anticipated so much of Descartes's argument: If I err, I exist—*Si fallor sum.*

With the thinking being's own existence comes the existence of its ideas: whatever they may be—worthy, silly, true, false, obscure, clear and distinct— their existence is part of what it means to be a thinking being. Notice that Descartes uses the word 'idea' in a sense as broad as that of the word 'thinking'. Thus, anything common sense associates with *sensation* (say, the color or fragrance of a piece of beeswax) is an idea, anything we can *imagine* is an idea, and anything we can *conceive of* (say, the idea of a triangle or the idea of extension in general) is also an idea.

We must therefore set aside much of what Plato meant by 'idea' and 'form' if we want to know what Descartes means by 'idea'. Recall that for Plato the mark of a Form/Idea was not its inherence in the existence of a thinking being, but rather its transcendence of any particular being, whether thinking or nonthinking. As it happens, in Meditation V Descartes distinguishes any *idea* he has of a triangle from what he calls there the *form, nature,* or *essence* of a triangle. The form/nature/essence of triangle is independent of his idea, although it is in some sense the cause of his idea. The form/nature/essence of triangle is thus a plausible candidate for the status of Platonic Form/Idea, but the ideas within Descartes's thinking being are not. Philosophical jargon permits us to ask just what ontological status those ideas have. Descartes's answer can be found in Meditation III. It is by no means a Platonic answer: he tells us that the ideas the thinking being finds within itself have their actual reality as *modes* of thought—that is, modes of thinking substance.

The Loss of the Commonsense World

In Meditation II Descartes uses the famous example of a piece of beeswax fresh from the hive to evaluate the ideas that are part of his existence as a thinking being. The example is a commonsense one, but the thinker is not at this point considering a piece of wax in a commonsense way: he has bought an ideal certainty at the expense of losing commonsense certainties. The thinker is not considering things but *ideas*—that is to say, ideas in Descartes's extended sense of that term: the still detectable sweetness of the honey and the fragrance of the flowers; the whiteness of the wax; its shape, size, hardness, coldness; the sound given off by it if you strike it; all these are ideas. Of course Descartes cannot draw attention to these ideas without speaking of the wax to which they purport to belong, but he does not mean to suggest that he knows the wax to be *there*, with an existence independent of himself as thinker.

He then notices that there is something unsatisfactory about these ideas: they are transitory and easily replaced by others; the thinker has only to heat the wax until it melts, and all the original ideas are gone. The hints of a taste of honey and the fragrance of flowers disappear; the color, shape, and magnitude change; the thinker is left with what common sense would call a hot and acrid liquid. But he is still not acknowledging common sense in a committed way, for what now exists for the thinker is merely a different set of ideas.

I said that Descartes wishes to evaluate various ideas and evaluate as well the functions of the thinking being that seem to be involved with the ideas. To do so he seems once more to invoke common sense. It is, he says, the same wax that remains; it is only the ideas that are evanescent, and so far all these ideas are of the kind that common sense would call sensations. Some ideas, however, have persisted through the transition—ideas of flexibility, changeability, and extension—and he can thus *imagine* that the same piece of wax can undergo all sorts of further modifications. For a moment Descartes's readers might suppose that imagining (which, like sensing, was on Descartes's list of functions of a thinking being) is a more reliable function than sensing. But Descartes immediately points out that the wax is capable of more changes than his imagination can possibly compass. It seems that only the bare idea of extension remains, together with whatever mental function he uses to attend to that idea.

Three conclusions emerge from the consideration of the piece of wax in Meditation II. First, the thinking being now has a clear and distinct idea of what common sense calls a piece of beeswax. It is the idea of extension, which includes the ideas of flexibility and changeability—includes, more precisely, the notion of the infinite transformability of one geometrical figure into another. This idea thus replaces the many transitory ideas that common sense would call sensations, and it replaces those ideas just because they are confused, while it is clear and distinct.

Second, the thinking being has established a rank among its functions: sensation has yielded to imagination, which in turn has yielded to the function that attends to the clear and distinct idea of extension. We might well call it the function of conceiving or understanding, and we might then call the idea of extension a conception rather than a sensation. Descartes, however, calls the function in question a *perceptio* or *inspectio mentis*, which translators commonly render as a 'perception' or 'inspection of the mind'. In the last paragraph of Meditation II, Descartes also remarks that the wax is properly perceived not by the senses or the imagination but only by the intellect (*solo intellectu percipi*).

Whatever we call that function, it is a close cousin to what Plato and Aristotle call by the name of *nous* and by other names from the same root. There is, however, one fatal difference: both Plato and Aristotle attribute a realistic competence to *nous* that Descartes does not attribute to *inspectio mentis*. *Nous* is said to be capable of knowing forms that have their being in independence of the mind that knows them, but Descartes's *inspectio mentis* deals only with ideas that are intrinsic to the thinking being.

Third, as Descartes points out towards the end of the second meditation, the most important outcome of the discussion of the piece of wax has been to confirm his own existence as a thinking being. To perceive the wax is only to *seem* to perceive it or to *think* that he sees it, for the thinker may in fact lack eyes—may indeed lack a body altogether. And although to think of the wax as extension (that is, to bring *inspectio mentis* into play) is to think of it more clearly than to think of it as a commonsense piece of wax, there is no assurance that extension exists.

At the end of Meditation II we have only a pure intellect or mind who is persuaded that *if he has a body set in a material world*, then the true nature of that body and that world is extension. In other words, if the physical world exists, it is what mathematics says it is and not what common sense tells us it is. Still, you might well wonder why in the world that mind/thinking being/intellect is troubled by all that muddle of what *appears* to be a commonsense world that contains such things as a piece of beeswax fresh from the hive.

The Cognitive Predicament Created by Descartes: Knowing (Representative) Ideas Rather than Things

Descartes's six meditations do not merely introduce his dualism; they also propound the central predicament of all post-Cartesian epistemology. The predicament, which begins to emerge in Meditation III, is this: ideas purport to represent real things or features of real things, but we have access only to the representations and not to their originals, if indeed there are originals. No one has presented that predicament more vividly than Locke, who accepts this part of Descartes's doctrine but rejects Descartes's rationalism:

> 'Tis evident, the Mind knows not Things immediately, but only by the intervention of the *Ideas* it has of them. *Our knowledge* therefore is *real,* only so far as there is a conformity between our *Ideas* and the reality of Things. But what shall be here the Criterion? How shall the Mind, when it perceives nothing but its own *Ideas,* know that they agree with Things themselves? (Locke 1690/1975, 563)

According to Descartes, ideas have not only *actual reality* as modes of the subject/substance called the mind, but also *representative reality.* He supposes that the thinking being knows only its own ideas directly, and that it can know other things—if in fact other things exist—only indirectly, by way of their representations in the ideas of the mind. Thus if subjects called bodies really exist—both the bodies that seem to make up the commonsense world and our own particular bodies, considered as parts of that world—we can know *what* bodies are (their true natures) and *whether* bodies really exist independently of mind only by undertaking a demonstration that begins with the representative reality of our own ideas.

The Attempt to Argue from Ideas to Things

In Meditation III Descartes tries to extricate himself from the predicament created by the nature of his initial certainties. He singles out one idea that has a privileged status, the idea of God understood as the Perfect Being. He argues that what is represented to the thinker in that idea assures the thinker of the existence of God. The chief resource used in the proof—I pass over the details of the proof itself—is a precursor of what is usually called the principle of sufficient reason, although it might with equal justice be called the principle of sufficient cause: "Now it is manifest by the natural light that there must be at least as much reality in the efficient and total cause as in the effect of that same cause" (para. 14).[1] In Descartes's causal principle, 'efficient cause' does not mean merely 'moving cause'; it also has a wider sense that originates in Christian monotheism: the sense in which a creative God can be regarded as efficient cause of the existence of the created thing.

It is of some historical importance that whatever plausibility the proof has depends upon the acceptance of this rationalist causal principle, and that the many empiricists who follow Descartes's account of knowledge in other respects do not accept the principle. Descartes, in any event, supposes that he has demonstrated the existence of the Perfect Being, and so at the end of Meditation III he has three certainties:

- that he himself exists as a thinking thing/being/substance (also called a mind or soul) in which various ideas exist as modes;
- that there exists an immaterial Perfect Being, or infinite substance, on which his own imperfect and finite being depends;
- that *if there should be a material world,* that world is not what it *purports* to be to what *purport* to be the philosopher's bodily senses but is rather extension in general—extension regarded as the spatial continuum contemplated by the mathematician, extension as known by *inspectio mentis* focused on an idea.

That last certainty is reinforced by Meditation V, which I here pass over except to say that it is concerned with the distinction between essence (*essentia/ forma/ natura*) and existence in the case of extended substance and the identity of essence and existence in the case of God, and that it provides an additional proof of the existence of God, the one called the ontological argument.

In any case, it is perfectly clear by the end of Meditation V that if extended substance does exist independently of the thinking substance's idea of it—if, that is, extended substance exists actually rather than merely as a representation in the mind—its existence must be quite distinct from that of mind. Mind can in principle exist apart from body, and body can have nothing in common with mind. It follows that there must be something of mere appearance in the commonsense version of the material world, the version Descartes must draw on to put forward the example of the piece of wax. The real physical world does not consist of many particular material substances but of one extended substance whose local configurations are as transitory as a piece of wax is.

I do not attempt to develop here the argument of Meditation VI, which purports to demonstrate that these quite distinct entities—mind and extended substance—exist together in the anomalous union that constitutes a human being. The intricate demonstration turns on Descartes's further scrutiny of his representative ideas and on the claim that the Perfect Being cannot be a deceiver. Intricate and ingenious as it is, I take it to be utterly unpersuasive. That we have bodies is true enough, to be sure; but we know that to be so on other grounds. The cognitive predicament established by Descartes is an unreal one; if we were indeed in such a predicament, there would be no way out of it.

An Absolute and Paradoxical Dualism

The chief point of this compact sketch of part of the *Meditations* is to make it clear that Descartes's dualism is both absolute and paradoxical: two things that are radically distinct are nonetheless intimately joined. In Meditation VI he calls this union of two substances, whose total independence of each other he has been at pains to establish, a *composite nature.* He even claims that the two substances joined in that nature interact, although in the *Meditations* he does not settle the question how interaction takes place but merely says that a particular part of the brain is responsible for it.

As early as the "Treatise on Man" he had singled out a certain "little gland" (the pineal gland) as being the focus of many of the brain's functions, but he does not explicitly make it the material locus of interaction, which is in any event not his concern in that work (Descartes 1991, 1:100). He takes that step much later, in his *Passions of the Soul,* which was written in the winter of 1645–46 but not published until late in 1649, just before Descartes's death. In articles 31–35 of that work, he claims that physical impulses are transmitted (however anomalously) to the mind by way of that little gland, and that the mind, which he supposes to be utterly immaterial, can also (equally anomalously) use that gland to produce changes in the (machine-like) physical world (ibid., 340–42).

The notion that the true nature of the physical world is pure mathematical extension has, as we saw, some precedent in Plato's *Timaeus,* and in some ways it has come into its own in twentieth-century physics. Descartes was no more an atomist than Plato was. Perhaps that claim should be qualified: although Descartes was not an atomist in the sense that generally prevailed before the present century, it could be argued that his interpretation of the fine-grained structure of physical reality in terms of vortices in extended substance was a better anticipation of today's views about atoms than was the notion of atoms held by the atomists of his time.

But all that is not our present concern. What can safely be said, however, is that whatever merits Descartes's image of the physical world as pure extension may have, it is hard to imagine constructing a *machine* on such a basis. Yet the image of the human body as a machine is, as we have seen, a recurrent one in his work.

Things Nature Teaches Us; Things Known by the Natural Light

Descartes nevertheless finds a place for the "knowledge" that he takes to consist of turbulent and confused *effects* in the mind of purely physical causes. He does so by making a distinction between what is true to a *mind* and what is given by God to the *composite nature* that is human nature—things given, as he says, "to me, as composed of mind and body" (*mihi, ut composito ex mente et corpore*) (Descartes 1641/1990, 202). Such things are in general useful to a being so composed: more often than not they will keep that being out of harm's way, which is to say that they tend to preserve its paradoxical union of soul and body. Descartes does not indeed say *"true* to me as a composite nature," as a twentieth-century linguistic relativist might say, but he comes very close to saying that.

His reason for not taking that final step may be seen in the distinction he makes in Meditations III and VI between what nature teaches us and what we know by the natural light. It is confusing that Descartes draws upon the word 'nature' to find names for both modes of apprehension, but the distinction is clear enough in spite of that.

- What *nature teaches* is practically useful, but its teaching is by way of confused ideas (like the sweetness and fragrance of honey) that are caused in the mind by what he merely calls in the *Meditations* a part of the brain. Such ideas represent the real physical world only in the sense that they are caused by that world, and caused in a temporally sequential sense that can easily be interpreted in terms of the received scientific doctrine of causality, which I discuss in Chapters 3 and 4. Being mere effects of their original, they *misrepresent* that original.
- What we know by the *natural light* must also come by way of ideas, for according to Descartes a finite mind can only know a reality by way of an idea that represents that reality. But the ideas associated with the natural light are clear and distinct, and they are not caused in the mind by way of a temporally sequential process but are rather innate in it.

Descartes here draws upon the Platonic tradition of *anamnēsis*, but with a difference. Platonic knowledge of the Forms/Ideas is not mediated by (Cartesian) ideas: it is a remembered direct *knowledge* of the Forms. The Platonic act of knowing is thus immediate in at least the sense that no idea that is ontologically a mode of the individual mind is needed to *represent* a known Form or Forms. Here we must not be distracted by the terminological difficulty I noticed in the previous chapter: for Plato, the Greek *idea* is an alternative to *eidos*, or Form. We can avoid the distraction by remembering that a Cartesian idea is not a Platonic *eidos/idea*.

We have met at least two examples of ideas that Descartes takes to be innate and thus known by the natural light. One of them is the idea of God, the Perfect Being; the other is the idea of extension. Descartes supposes that his ratio-nalist causal principle—the one I called a precursor of Leibniz's principle of sufficient reason—is also known by the natural light. We can make this suppo-sition consistent with the cognitive importance Descartes gives ideas only by supposing that a finite mind knows that principle by way of a cluster of ideas that represent the real principle, which Descartes obviously supposes to have a reality independent of our formulation of it.

To call an idea innate in a Cartesian setting is by no means to say the idea is uncaused, although it clearly cannot be caused by a temporally sequential series of events in the brain. The mode of causality Descartes envisions is non-temporal, nonsequential: innate ideas are caused in the created mind by the same nontemporal act of God that creates the mind. Recall, however, that a Cartesian idea has two modes of reality: its actual reality as a mode of thought and the representative reality by virtue of which it represents something other than itself. Descartes takes it for granted that the (eminent) efficient causality of God accounts for both modes of reality.

The Influence of Descartes's Dualism

Neither the rationalists nor the empiricists who come after Descartes accept the notion that thinking substance and extended substance can coexist in a composite nature in which they interact. The great rationalists argue that mind and body are related in an attunement of a nontemporal kind, and they make God responsible for this attunement—a parallelism in the case of Spinoza, a preestablished harmony in the case of Leibniz. To put forward their own solu-tions they must also make major adjustments in the Cartesian notions of God and substance.

The empiricists who shaped the empiricism of the twentieth century tend to ally themselves with the sciences, and indeed major figures like Locke and Hume thought of their investigations of human understanding as attempts at a science of human nature. Dualism dies hard, as hard as Western Christianity itself, and it died hard in empiricism; but in that setting it was never a dualism as extreme and clear-cut as that of Descartes. Yet it seems fair to say that the long-term tendency of empiricism is towards the monism called materialism.

No doubt one reason for that tendency was the acceptance by empiricism of one major feature of Cartesian dualism: the claim that the reality of the physi-cal world is different from what common sense takes it to be. It is fundamental to empiricism that the commonsense world which purports to contain things like Descartes's famous piece of beeswax is an appearance, and that science is more competent than common sense or indeed philosophy to give a true account of the reality behind the appearance.

But if Descartes's dualism and some of his views on knowledge were gradually rejected by the parallel development of empiricist philosophy and science, enough of Descartes's views were incorporated into the empiricists' theories of knowledge to leave them with a formidable problem about the cognitive accessibility of the material world. Descartes himself did not have that problem. Right or wrong, he was a rationalist, and the principle of sufficient reason and the doctrine that the mind is a substance capable of receiving innate (though representative) ideas provided him, as he thought, with the possibility of penetrating the veil of appearance to find the reality behind it.

But the empiricists rejected that part of Descartes, and indeed in due course they rejected the notion of substance altogether. Nevertheless, they accepted Descartes's claim that knowledge does not consist in a relation between our mind and things but rather in the relation between the mind and its ideas—ideas in Descartes's inclusive sense. The subclass of Descartes's ideas that consists of what are more commonly called impressions was soon singled out from that more inclusive class of ideas and identified as the real source of knowledge. But their status as such a source was at once undermined, for impressions were no longer regarded as having a representative value from which inferences could be made to realities that were not impressions. The natural tendency of developed empiricism is thus towards phenomenalism—a stream of experience consisting of impressions whose causes, if any, are unfathomable, and whose chief cognitive significance is that it is prudent to anticipate their future course insofar as that is possible. If the material world as described by scientists is the reality behind the experienced impressions, there is no way of demonstrating that to be so on the basis of that experience.

This development is one, but by no means the only, source of what I called in the Introduction the negative philosophical judgment about the power of mind. It has not hindered in the least the dominating growth of materialism, but it has left empiricist philosophy with a profound appearance-reality problem that has never been satisfactorily resolved. Empiricism, having set out to distinguish reality from appearance, having made the knower itself (mind itself) a kind of appearance, finds itself towards the end of our century thoroughly entangled in a network of appearance of its own contriving. It longs for realism, but is unwilling to take the radical step that would reclaim our natural dominion in the real.

A related development of which Descartes's dualism has been an indirect cause has led to an assumption that has profoundly distorted the study of the mind in twentieth-century empiricist philosophy. It is the assumption that mind and body may be understood in terms of two distinct streams of states/events, one mental and the other physical.

The origin of the notion of a succession of mental states/events deserves some comment. We might suppose the notion to be implicit in Descartes's dualism, but it is not explicitly developed there. For all practical purposes, it is developed in detail in the parallelism of Spinoza, which recognizes a kind of distinctness of mind and body but allows for no interaction between them. I say "for all practical purposes" because Spinoza does not use the expressions 'mental state' and 'mental event', which are more recent. Spinoza uses much of Descartes's terminology (for example, 'substance', 'attribute', 'mode'), and he is obviously influenced by Descartes's dualism, but his doctrine is a monism; for there is only one substance, and mind and body are merely two of its attributes. As we have seen, the notion of substance is so generalized in Descartes's hands as to bear little relation to what Aristotle meant by *ousia*. In Spinoza's system, the notion of substance becomes completely generalized: in effect substance is equated with Being.

For Spinoza, that one substance is identical with the whole of nature and with God as well: 'nature', 'God', and 'substance' are thus quite interchangeable terms. Mind (in general) and body (in general) are two of an infinitude of *attributes* possessed by God/substance/nature. Particular things are merely particular *modes* of substance. A human being (having a particular mind and a particular body) is merely one of an infinitude of modes of substance. But human beings are not unique in existing in both the attribute of mind and the attribute of body: all modes, including those that are not sentient, exist in both attributes—and indeed in other attributes of which we know nothing.

Causality, according to Spinoza, operates with necessity: substance is pervaded by a causal necessity, which manifests itself in the sequence of ideas (and other mental things) and in the sequence of physical motions. But it manifests itself in that temporal order only because of the nontemporal causal dependence of modes and attributes on the one substance in which they inhere. That one substance is causally dependent on nothing outside itself; indeed there *is* nothing outside it. God/substance/nature necessarily exists, and for that reason Spinoza says that it is self-caused (*causa sui*).

Nothing in the attribute of mind causally influences anything in the attribute of body; nothing in the attribute of body causally influences anything in the attribute of mind. Yet the two attributes are absolutely parallel, and for everything in either attribute there is a counterpart in the other. All this goes back to the underlying causal necessity of God/substance/nature—a necessity difficult to distinguish from *logical* necessity. Spinoza nevertheless makes morality the center of his philosophy: his major book, which lays out systematically and exhaustively what I have merely sketched here, is in fact called the *Ethics*. To propound his moral philosophy he makes creative and ingenious use of the principle of parallelism: to be in bondage to one's passions has as its mental parallel a confusion and lack of distinctness in one's ideas; getting one's thoughts clear and distinct (as Descartes recommends) has as its parallel the control of one's (bodily) passions.

But Spinoza has not been a major influence on the empiricist tradition. We must look to Hume if we are to understand the empiricist notion that mind consists of a sequence of what today would be called mental states/events. In Hume's language, our experience consists of a sequence of impressions both external (color, for instance) and internal (hunger, for instance), and our ideas are but faded copies of the original impressions. The distinction between external and internal is not based on an appeal to common sense: it is simply part of the impression of a particular color that the color is external and part of the impression of hunger that it is internal to oneself. The existence of the external world is no more taken for granted by Hume than it was by Descartes; but, unlike Descartes, Hume argues that we cannot demonstrate its existence (Hume 1748/1955).

Since from the phenomenalist perspective of Hume the mental stream of states/events is more certain than the physical stream, empiricism is left with a difficult problem. On the one hand, it tends to regard the stream of physical states/events as more fundamental than the mental stream, and it wishes to regard the physical stream as pervaded by the kind of causality I discuss in Chapters 3 and 4 under the rubric of the received scientific doctrine of causality. On the other hand, it must regard the existence of the external physical world as a belief, postulation, or construction on the basis of the mental stream; and it must derive its notion of causality not from a cognitive encounter with necessities in the external world but from the mere regularity of the experienced sequence that is the mental stream.

If Hume is strictly followed, the causality attributed to either the mental or the physical stream can mean no more than constant correlation: x is the cause of y if and only if when x occurs y follows and when x does not occur y does not occur. Moreover, any causality attributed to the postulated physical stream must be a version of the causality of the stream of impressions. Those empiricists who take both the mental stream and the physical stream for granted must confront a profound problem about which stream has ontological priority and about the causality (if any) attributed to either stream. The normal empiricist inclination towards materialism is countered by the phenomenalism discussed above and by the felt need to employ Hume's diminished sense of causality, which is associated with his phenomenalism, in dealing with the material world.

The obsession with the notion of parallel streams of mental states/events and physical states/events, each arranged in what at least purports to be a significant causal series, has a significance for the study of mind that ranges well beyond the problem I have just touched upon. I take up that matter towards the end of Chapter 4.

3

The Received Scientific
Doctrine of Causality

Two Conflicting Attitudes towards the Power of the Human Mind

D O C T R I N E S   A R E   made by the mind, partly in response to natural concrete things, which are not made by mind, and partly in response to other doctrines. This book is primarily directed towards the concrete, towards the actual, so I have pared my references to doctrines to a minimum so far. But the received scientific doctrine of causality deserves special attention, for its prestige prevents many professional philosophers—and many persons in other fields who study the mind in their own ways—from looking to see whether the concrete world does or does not exhibit hierarchical causality.

Two conflicting attitudes preside over the development of the scientific doctrine of causality from the beginnings of modern science in the Renaissance down into our own times. The first is a confidence in the power of the human mind to know, though only after hard work, the nature of things. By the time Newton's *Principia*—published in 1687—had been absorbed by the intellectual community, that confidence was considerable; since then it has increased at equal pace with the exponential growth of science and technology, so much so that it may now be called a superb confidence; and although it sometimes mounts into hubris, it is for the most part justified. Among the writers responsible for the development of the received scientific doctrine, Laplace stands out as the most important representative of this attitude.

The second attitude is one of diffidence about the power of mind to know the nature of things. The more remote origin of the attitude lies in certain of

Descartes's doctrines I discussed in the preceding chapter, although Descartes's own confidence in the power of the mind does not seem to have been affected by those doctrines. Great philosophers sometimes fail to draw the consequences from their doctrines which are in fact there to be drawn. The nearer origins of the attitude lie in the theory of knowledge that comes down to us from philosophical empiricism and the various attempts to amend or circumvent the findings of empiricism; it culminates in a negative philosophical judgment about the power of human knowing to get at the nature of things—if indeed there is a nature of things that is independent of the knower. Hume is the most influential representative of this attitude among the writers responsible for the received scientific doctrine.

There is thus a disturbing tension at the heart of the scientific doctrine of causality as we have it today, a tension that has only increased with the years. Although it is clear enough how the judgment came about, it is nevertheless paradoxical, for the original impulse behind empiricism came from a profound respect for the early achievements of Western science, a respect that led to an attempt to extend the scope of science to human nature in general and to human knowledge in particular. The attempt persists today in many efforts to "naturalize" epistemology and many efforts to characterize the mind-body problem as just another scientific problem, although a highly complex one.

Transformation of the Fourfold Scheme of Causality by the Doctrine That God Created the World

The hierarchical doctrine of causality received from antiquity recognized four causes—four modes of explanation—that were required to explain any entity or event. We owe their dominant names to Aristotle, but from Chapter 1 it should be clear that each of these causes has its precedent in Plato. Aristotle's names for at least three of the four causes are anthropomorphic and down to earth in the Greek, but our own translations have by now acquired a technical and abstract air that does not quite do justice to the original.

The *material cause*, in the simplest case, is whatever things a complex entity is made of—the elements, components, or ingredients which, in their coming together by virtue of a complex series of motions, produce the complex entity. The *moving cause*, or *efficient cause*, as it is more often called, is the motion or complex of motions by which the complex entity is generated (in the case of natural things) or manufactured (in the case of artifacts). There is a certain relativity to material and efficient causes, as there must always be in a hierarchical doctrine: if we are not satisfied with a certain account of a material cause, we may go on to find out the material causes of whatever material cause we began with; if we are not satisfied with a merely local version of efficient causality, we may go on to look for earlier or more fundamental motions (or sources of motion) that contribute to whatever we are trying to explain.

The *formal cause* is in one sense the visible structure and in another sense the intelligible structure of the complex entity that emerges from the process. Because form is regarded as intelligible, it is also universal: the same form can be shared by many animals, many artifacts. The *telic cause,* which is more commonly known by the Latin-based translation *final cause,* is the purpose or end subserved by the process that brings about the complex entity. It is obvious that many telic causes can operate in any one instance. On the other hand, in one subtle sense the formal cause and the telic cause can coincide: the completed and unchanging form is precisely what the process, whether artificial or natural, aims at. Indeed, as we saw in Chapter 1, Aristotle fuses the efficient, formal, and telic causes in the case of the soul. Our theme in that chapter was the human soul, and I made the point that for Artistotle the soul is the cause, in all three of those senses, of the living body—and indeed of the entire functioning—of the rational animal. But of course Artistotle's claim extends to all living creatures: the soul, of whatever kind, is a cause in that threefold sense.

This well-known view of causality is complicated by the monotheistic tendency in Greek philosophy. In the case of Plato the causal and moral role played by the Form of the Good is clearly monotheistic, and so is the great myth of the Demiurge in his *Timaeus.* On the other hand, most of Plato's myths are couched in conventional polytheistic terms. If we assume monotheism to be his most considered view, as I think we can, then we may say that Plato's God, whether it is the Form of the Good alone or an agent-like principle qualified by the Form of the Good, exercises a causality that includes form, purpose, and source of motion—in short, what Aristotle means by formal, telic (final), and efficient causality. In any event, Aristotle's God, which he characterizes in several ways—the Unmoved Mover, thought that thinks about thinking, pure form, pure actuality—is quite clearly the ultimate formal, final, and efficient cause of whatever happens in the universe. Aristotle's God does not of course move the world by imparting movement to it; he is the efficient cause of movement only in the sense that he is the unmoving actuality (form) towards which all finite actualizings tend as towards a goal or an object of love.

But for Plato and Aristotle God does not create the world. For Plato, at least in the mythical language of his *Timaeus,* an order-providing, agent-like figure is confronted by an uncreated principle that is close to not-being, a principle so close to nothing at all that it almost escapes our rational grasp. He calls that principle the Receptacle—an utterly empty space that lacks any inherent order of its own beyond what that vague image suggests. In the less poetic language of Aristotle, the same principle is called prime matter: it is something upon which the unmoved mover has an ordering effect, although not an agent-like one.

Such makers of the modern scientific era as Copernicus, Kepler, Galileo, and Descartes inherited this complex doctrine of causality, but only after it had passed through a profound modification by the three great monotheistic religions which shaped the philosophy of the high Middle Ages. In that period all four of the Aristotelian causes, or modes of explanation, were gathered up in the notion of a personal God who created and sustained the world—a God whose proper philosophical name was Being, and sometimes even the *act* of Being. One important consequence of this development was the attribution of the very existence of prime matter (the Platonic Receptacle) to the creative power of God. With that change, matter in the relative sense could still enter into a causal explanation, but prime matter could no longer be regarded as an ultimate causal principle.

This change in the notion of the material cause was made possible by an even more momentous transformation of the notion of the efficient cause. In Aristotle the efficient cause was the *moving* cause—either a movement transmitted from one thing to another or an ultimate source of motion—but the notion of the efficient cause now came to include also the creative source of the very being/existence of any being that was capable of motion. The use of the Latin-based expression 'efficient cause' to translate Aristotle's Greek for 'moving cause' helped to make this transformation plausible, for 'efficient' goes back to the Latin root for making, doing, acting.

The modifications in the inherited causal scheme were more thoroughgoing than these examples suggest. The superabundant creative cause was now regarded as, so to speak, the cause of any causality—formal, final, efficient, or material—exercised by any finite being. In this guise the first cause was sometimes said to possess *eminent* causality. It is a commonplace of medieval philosophy that God as First Cause or creator of the universe is also its ruler and law-giver. Everything was thus already in place for the doctrine that the physical world is pervaded by laws of nature laid down by a rational and provident God—the doctrine, in fact, that dominated scientific thinking from its beginnings in the Renaissance until the end of the eighteenth century.

But that doctrine does not yet appear in characteristically medieval philosophy. In that setting the appropriate goal of human life was assumed to be eternal salvation, and the doctrine of natural law was confined to moral, legal, and theological matters. The notion of what we today call the laws of nature had not yet arisen, although the old moral notion of natural law probably helped prepare the way for it.

God as the Source of Mathematical Laws of Nature: The Divine Reductionist

The conviction that the physical world is governed by mathematical laws of nature that originate in the mind of God is adumbrated in the earliest speculation of the young Kepler in the 1590s, given its first respectable scientific form in his enunciation of his three great laws of planetary motion—the first two in 1609, the third in 1619—and confirmed in some detail by the subsequent work of Galileo. By Galileo's death—in 1642, the year of Newton's birth—it had become the orthodoxy of the new science, and it remained so through most of the eighteenth century. Although I have been using the Aristotelian causal scheme as a convenient guide, Kepler stands more in the tradition of Plato—especially that of Plato's *Timaeus*, in which the influence of Pythagoras is most vivid—and also of Plotinus, whose imaginative Neoplatonism did so much to further the tendency of medieval monotheism to locate the Platonic Forms in the mind of the creator.

Kepler thus had a near precedent, in the Demiurge of *Timaeus*, for his doctrine that the creator endows the physical world with mathematical harmonies that can be found by the investigator (Plato, *Timaeus* 35B–36D, 38C–39D, 53C–57D). The Demiurge, to be sure, is not a creator but a divine artisan or constructor (as the Greek word says), who works with the Receptacle. But E. A. Burtt (1932, 53) is surely right to say that Kepler's work tended to eliminate Aristotle's formal causality from physical science and to replace it with a kind of mathematical formal causality, and that Galileo's work tended to set God's purposes outside the business of physical science and thus eliminate final causality from consideration by the scientist (90). It would be unhistorical to assume that Galileo thought that God had no purpose in creating the world: his point was rather that whatever purpose there might be was a matter for religion rather than science.

Kepler believed that the harmony expressed in a mathematical law is the reason for and thus the cause of a physical process no less than the identifiable physical factors or entities that are involved in that process. For Kepler, however, as for Plato in *Timaeus*, physical factors identifiable in mathematical terms are more important for a scientific account than items identifiable as what Aristotle would have called *ousiai* (entities/beings). There is thus visible, even at this time so early in the development of science, a tendency to resolve commonsense entities into factors that can be handled mathematically. This reductive tendency was soon augmented by a quite different way of reducing commonsense entities to something alleged to be more basic: atomism, which flourished as an alternative philosophy even in the time of Plato, was now revived and soon became a part of the interpretive scheme of the most influential of the new scientists.

The acceptance of atomism by Galileo introduced into precise science not only the distinction between primary and secondary qualities we have come to associate with science but also the possibility of regarding commonsense entities themselves as ontologically secondary when contrasted with the primary ontological status of the atoms. The two reductive tendencies thus introduced so early—resolution of commonsense entities into mathematically expressed *features* on the one hand and into more basic *entities* on the other—are still with us in today's science.

It would seem, however, that in today's versions of reductionism the emphasis on (general) features is the dominant one: thus, although it is often supposed that commonsense entities can be reduced to basic microentities, the microentities themselves are brought into equations in terms of general predicates (features), such as mass, charge, velocity, electromagnetic field, wave-length, and spin. Any microentity, such as an electron, is in effect regarded as a mere x characterized by those predicates. Most of the materials of twentieth-century reductionism were thus early to hand in the development of science. The word 'reduction' is used in other senses as well: we shall meet later a different sense that has been of great importance for twentieth-century philosophy.

Summary of the Received Doctrine as It Was in the Eighteenth Century

The doctrine of scientific causality that was in place by the middle of the seventeenth century—although by no means fully explicit—was still there, in more developed form, in the eighteenth century. It is profitable to sum up the doctrine in terms of the ancient fourfold causal scheme of Aristotle, if only to see what a profound modification of the ancient scheme had taken place.

- The interactive movements of observed macroscopic entities such as the celestial and terrestrial objects investigated by Newton are instances of *efficient (moving) causality*. So also are the inferred movements of the inferred microentities called atoms; indeed, the observed macroscopic entities move as they do move because they are structures made up of such microentities.
- The inferred microentities are thus *material causes* of observed macroentities, but only in a diminished sense of Aristotle's notion of material cause, because correlative formal causes (in Aristotle's sense) were no longer recognized within science. To express the point in another way: the new science does not recognize the hierarchical principle by virtue of which what was a formal cause at one ontological level could serve as a material cause for a higher (formal) ontological level. Indeed, the new science inverts the hierarchical principle, for the microentities, though

material causes in the diminished sense, are regarded as more truly real than the entities to whose macroscopic structure they contribute. There is therefore something not quite authentic about the observed forms of macroscopic entities, since these forms are dependent on the observer in a way analogous to such secondary qualities as colors.

- The laws of nature discovered in the course of the seventeenth and eighteenth centuries—they are essentially the laws of Newtonian mechanics, which had been inferred from the observation of macroscopic objects—provide a diminished version of Aristotelian *formal causality* in the sense that they display the mathematical form in accordance with which atoms, and so also macroscopic entities, move. Other laws remained undiscovered in the eighteenth century, laws about the structure of microentities such as atoms and so about the possible macrostructures that arise when microentities come together and cohere; but it was taken for granted that these laws too would turn out to be formal only in the mathematical sense of the laws already discovered.

- The absence of formal causality in the full Aristotelian sense—a sense that includes biological form—means that the scientific doctrine of causality recognizes no *telic (final) causality.*

To say that this doctrine of causality dominates eighteenth-century science is not to say that it dominates eighteenth-century thought in general. Leibniz's great attempt to reconcile mechanism and teleology culminates in the first few years of the century, and its spirit can still be seen in Kant's attempt towards the end of the century to overcome both unqualified empiricism and unqualified mechanism. It is a complex century, and in some ways its thought is markedly theological and therefore markedly teleological. Newton, for instance, after completing the *Principia* in 1687, when he was in his forty-fifth year, devoted as much time to theology as to science during the rest of his life. He was not alone in taking theology seriously: for most educated people the laws of nature were still generally regarded as God's laws—laws that are as they are because they subserve God's purposes even though the laws as expressed in mathematical terms are not telic. Indeed, it was the very success of science that led so many eighteenth-century writers to think of God as the Divine Artificer.

The Simplified Causality of Laplacean Determinism: The Lawful Succession of the States of a Physical System

The determinism implicit in the scientific doctrine of causality haunted some of the best minds of the eighteenth century. The motion of a body—whether microscopic or macroscopic—in accordance with the laws of nature was thought to necessitate the movement of another body with which it collided and to necessitate also the equal and opposite reaction upon itself that results from the collision; and what was true of any investigated instance appeared to be true also of any collection of bodies, however large or complex. The determinism was mitigated during much of the century by the apparent failure of Newton's laws to account for certain irregularities of planetary motion. It was Newton's own view that the creator occasionally intervened in his created system to adjust what would otherwise be the cumulative effect of such aberrations.

But beginning about 1773, Laplace was able to provide mathematical demonstrations that the supposed irregularities in the planetary motions—especially the incommensurability of the motions of Jupiter and Saturn and certain aberrations in the motion of our own moon—were accounted for by long-term cyclical regularities, that is, by the invariability of planetary mean motions. The detailed summary of his achievement, which took many years and drew heavily on the work of Lagrange and Legendre, is given in his great *Mécanique céleste*, published in five parts between 1797 and 1825.

Before the end of the eighteenth century, the supposed role of a divine rational agent in establishing the laws of nature was therefore much eroded: Laplace's famous response to Napoleon's question about the role of God in his system, a system welcomed in its time as the completion of the achievement of "the incomparable Mr. Newton," is a convenient cultural mark for the end of the dominance of the notion of a divine lawgiver: "Sire, I have no need of that hypothesis." But it is by no means clear that Laplace himself rejected the notion of a divine lawgiver. He had no need of the hypothesis in this sense, that a body of eternal and prescriptive laws of nature seemed to him to be coincidental with the nature of the material universe studied by the scientist, so that to understand the material universe was to understand just those laws.

To put the matter differently, the laws cannot be other than they are if the material universe is what it is. If the universe is indeed created by a divine intelligence, that intelligence's very idea of the material universe it wishes to create carries with it just those laws. In that sense the laws are necessary rather than contingent: the creation of the material universe is necessarily the creation of the laws that permeate it and define it. Many of Laplace's contemporaries took him to be an atheist, but it is risky to attribute an unqualified atheism to him: his private correspondence is said to contain remarks inconsistent with atheism, and his vision of the universe as a perfectly ordered system permeated by law—a system that requires no tinkering on the part of an intervening God— is consistent with the common eighteenth-century notion of God as the Divine Artificer. But at least it seems clear Laplace did not suppose that the question whether the universe was created by a God was relevant to his scientific enterprise.

Laplace's well-known determinism is crucial to our understanding of the scientific doctrine of causality, so it is important to notice that the determinism does not seem to have been a conclusion he drew from the success of the investigations that led to the *Mécanique céleste*. The passage so often quoted to illustrate his determinism—I quote it myself in due course—occurs in a philosophical essay on probability that was first published in 1814, at a time when his work on the *Mécanique céleste* was substantially complete, but Roger Hahn has demonstrated that Laplace's determinism was in effect a premise of the great research program that produced the *Mécanique céleste* rather than a conclusion drawn from it (Hahn 1967, 14–18).

Early in Laplace's career Alembert wrote a paper in which he discussed the possibility that some of the laws of nature are necessary and thus could not be other than they are, while some are contingent in the sense that they depend only on the arbitrary will of the creator; Alembert touched also on the question of determinism but of course could take no decisive position on it. Somewhat later Condorcet discussed the same issues and expressed a hypothetical form of determinism. Laplace responded decisively in 1773 with an unconditional determinism, together with the corollary claim that the laws of nature are in fact necessary. I do not repeat here the 1773 passage quoted by Hahn: it is enough to say that the passage anticipates in all essentials Laplace's well-known 1814 remarks. It seems clear, then, that the 1814 remarks should be taken as a more mature statement of a metaphysical point of view which Laplace had adopted at the outset of his career and which then guided the whole of his scientific life.

I give the 1814 remarks in a more complete form than is usual because the passage so often quoted is preceded by two passages of some importance. In one of them Laplace rejects the existence of chance except in a sense that is merely a measure of what we do not yet know about the universe: the gradual overcoming of our partial ignorance is the gradual reduction of what we ascribe to chance. In the other he shows his utter confidence in the powers of the mind by relating this conclusion to the principle of sufficient reason.

> All events, even those which because of their insignificance do not seem to abide by the great laws of nature, are a consequence of them just as necessary as are the revolutions of the sun. In ignorance of the ties that unite those events to the entire system of the world, we have made them depend upon final causes or upon chance, according as they occurred and followed one another with regularity or without apparent order. But these imaginary

causes have been successively put back with the boundaries of our knowledge, and they disappear entirely before sound philosophy, which sees in them only our ignorance of the true causes.

Present events have a connection with the preceding ones founded on the self-evident principle that a thing cannot come into being without a cause that produces it. This axiom, known by the name *the principle of sufficient reason,* applies even to actions that are judged indifferent. Even the freest will cannot bring such actions to birth without a determining motive; for if, in two situations whose circumstances were exactly the same, the will were to act in one and abstain from action in the other, its choice would be an effect without a cause: it would then be, says Leibniz, a case of the blind chance of the Epicureans. The contrary opinion [contrary, that is, to the view that the will cannot act without a determining motive] is an illusion of the mind which, losing sight of the fleeting reasons operative in the choice of the will in indifferent things, persuades itself that the will determined itself by itself and without motives.

The second paragraph ends in an excessively compressed reference to Leibniz's discussion of free will, so I interrupt the long passage here with what I hope is a clarification. Laplace is insisting, as Leibniz also did, that the principle of sufficient reason—a thing cannot come into being without a cause/reason that produces it—is without exception. Even the process of willing is subject to it, though it sometimes seems otherwise: although some proposed actions are thought to be indifferent (no motive whatever inclining the will either to act or to abstain from action), there are in fact no such actions. The Epicurean version of ancient atomism held that atoms sometimes swerved in their courses without any reason/cause for their swerving. This doctrine of blind chance—that is, radical or absolute chance—thus provides the possibility that some "choices" are absolutely without determining motive; and of course it also provides for exceptions to the principle of sufficient reason. But both Leibniz and Laplace reject the doctrine of the swerve of the atoms—the *clinamen atomorum* as Lucretius called it. When Laplace speaks of chance, as he does throughout the treatise from which the quotation is taken, he does not mean radical or absolute chance: he means relative chance—chance that is a function of our ignorance. The passage now continues:

We ought, then, to regard the present state of the universe as the effect of its prior state and as the cause of the state that is to follow. An intelligence which could know all the forces animating nature at a given instant and the corresponding positions of all the beings of which nature is composed—which, moreover, was vast enough to submit all these data to analysis—would embrace in the same formula the movements of the largest bodies in the universe and those of the lightest atom: nothing would be uncertain for such an intelligence, and the future, like the past, would be present to its eyes. The human mind offers, in the perfection it has been able to give to astronomy, a rough draft of this intelligence. Its discoveries in mechanics and in geometry, coupled with that of universal gravitation, have put it within reach of comprehending in the same analytical expressions both the past and the future states of the system of the world. By applying the same method to certain other objects of its knowledge, it has succeeded in bringing observed phenomena under general laws and in anticipating those phenomena that given circumstances ought to produce. All these efforts in the search for truth tend to bring it incessantly closer to the vast intelligence we have just imagined, but from which it will always remain infinitely distant. (Laplace 1814/1921, 1:2–4)

Note that Laplace wants his doctrine to apply to the "lightest atom" no less than to what we today tend to call macroscopic things. Supplementing his scientific achievement in celestial mechanics by the imaginative picture provided by the renewed interest in ancient atomism—in view of Newton's acceptance of atomism, it might be more accurate to speak of a general *commitment* to atomism on the part of scientists—Laplace has made an imaginative leap to what he supposes nature really is as a concrete physical system: a collection of microscopic particles moving inexorably from one state to another and giving rise to all the macroscopic realities to which human beings respond.

Laplace's doctrine of causality comprises two interdependent notions: (1) the universe as a whole is a physical system that passes through successive *states*, any given state being the *cause* of the state that follows; (2) the transition from state to state is governed by laws of nature—that is to say, the laws of nature are causal factors no less than the physical states are. In practice, of course, scientists must deal with physical systems smaller than the entire universe. We therefore arrive at this working model for Laplace's doctrine of causality, in which *PS* stands for some (concrete) physical system, *s1* stands for a state of that system at a given time and *s2* for a state at a later time, and *L* stands for some law or laws of nature.

$$\overbrace{PS(s1) \rightarrow PS(s2)}^{L}$$

Notice that a physical system, suitably defined at a certain time, is regarded as the cause, and *that same system* at a later time is regarded as the effect. In the case of the comet fragments that struck Jupiter in 1996, the quite large system including both the comet fragments and Jupiter a minute or two before the first impact would be the cause, and the same system just after the last impact would be the effect. The notion of a state of a physical system has become an important part of twentieth-century philosophy of science, and various rigorous mathematical definitions of such a state have been provided. As for the laws of nature, philosophers today often say that they *cover*, or *back*, the transition from a $PS(s1)$ to a $PS(s2)$. Such noncommittal language suggests that the status of the laws of nature remains an open question. Keeping in mind twentieth-century refinements and twentieth-century open questions, we may safely use the above model to discuss the received scientific doctrine as it is today.

It bears repeating that the attitude towards the human mind expressed in the passages quoted from Laplace is clearly the one I called earlier the confident attitude. Although Laplace distinguishes the human mind from the God-like intelligence it tends towards as a limit, he takes the power our mind possesses right now to be considerable, as his invocation of the great rationalist principle of sufficient reason assures us. He supposes that there really is a concrete universe independent of our finite mind, a universe that moves inexorably in continuous becoming in which any arbitrarily chosen state is the cause of a later one (and an effect of an earlier one). He also supposes that a body of mathematical theory such as the one developed in his *Mécanique céleste* captures with remarkable faithfulness all the sinuosities of that concrete becoming.

From the empiricist point of view, which leads eventually to the negative philosophical judgment about the power of the mind, there are difficulties about the conjunction of those two suppositions. The suppositions require that we recognize two senses of the notion of a physical system, one concrete and the other abstract, so let us now use two abbreviations: PS for a concrete physical system and PS' for an abstract physical system—one expressed, as the *Mécanique céleste* is, in terms of a body of mathematical theory. The difficulties, then, are these: How can we assure ourselves of the independent reality of the concrete physical system (PS)? Supposing that there is indeed an independent PS, how can we assure ourselves of the faithfulness of the PS' to that PS? For twentieth-century versions of Hume's empiricism, no PS, macroscopic or microscopic, can be known as such; and any (abstract) PS' can be no more than a rational construct. How can a cognitively inaccessible PS be a basis for a PS'? How, on the other hand, can a PS' enlighten us about such a PS?

Twentieth-century philosophy is full of such problems, so much so that some influential philosophers of science take it for granted that when you talk about a physical system, you ought to mean only an abstract, ideal physical system, that is, a *PS'*, for although they concede that we do have access to empirical stimuli (sometimes called phenomena), they think we do not have direct cognitive access to a concrete *PS* as such (Suppe 1974, 224). In this setting, obviously, Laplace's determinism must be modified: because of the impossibility of establishing complete parameters for the *PS*, and because of the approximate character of the parameters we do in fact establish for a measured first state and second state, we can only *impute* determinism to the concrete *PS* that is the subject of the investigation. It is the (abstracted/constructed) *PS'* that is truly deterministic, and it is so because it is a thing of *mathematical* necessities. The "movement" from *s1* to *s2* is not a physical one in the case of the *PS'* but a mathematical necessitation.

Laplace's confidence in the power of the mind, however, seems to have been such that he was quite untroubled by such difficulties, even though Hume's philosophical work was well known in France early in Laplace's scientific career. Let us therefore set such difficulties aside for the moment and return to them when we consider the second attitude towards the power of mind—the diffident one.

Meanwhile, we may safely take Laplace as the prime exemplar of the working scientist's confidence in the power of human reason. Building upon Newton's work, he resolves planets and other macroscopic objects into such general features as position, mass, velocity, and acceleration, and he derives particular quantitative values from whatever concrete physical system he is dealing with. These parameters of that *PS*—today they would be called state variables—are then used to define the first state of an abstract, or theoretical, physical system (*PS'*). These parameters, that is to say, become the variables for differential equations based on the laws of nature that are thought to prevail in the concrete *PS*. The solutions of the equations then yield mathematically a certain other state of the (abstract) *PS'*, defined by other values of those same parameters. These parameters are then construed as predictions about the concrete *PS*, and so the familiar circle is closed. If the work has been sound, these inferred state variables will yield sound predictions about the *s2* of the *PS*.

The definition of a state varies considerably today for such different fields as mechanics, electromagnetism, and quantum theory; so also do the equations whose solutions yield us the *PS'*(*s2*) on the basis of a *PS'*(*s1*) (Nagel 1961, 279–93). But the success with which this circle is closed is as characteristic of confident twentieth-century practice as it was of Laplace's practice in a simpler situation.

The notion of a concrete physical system is philosophically unproblematic when the attitude of confidence in the power of the mind prevails. The system might be a group of comet fragments moving towards Jupiter, or it might be a part of a brain. Such systems are not easy to know: the comet fragments and Jupiter are, after all, very remote; the live brain is extremely complicated and difficult of access. But if the confident attitude prevails, physical systems are not regarded as mere postulations arising in response to the physical stimulation of our senses on the one hand and scientific theories on the other hand. We need not, that is, give up our confidence that our reason is capable of expressing in a *PS'* something true about a concrete physical system that is truly independent of the knower—a *PS* that is truly *there* to be known.

The notion of a (concrete) physical system is a relative one: any *PS* we choose to investigate will be enframed by a larger *PS* and will itself enframe smaller physical systems. Perhaps there are upward and downward limits, perhaps not. In a literal sense an enframing *PS* is the boundary of an enframed *PS* and so establishes certain conditions that must be taken into account in studying the enframed *PS* with the help of a constructed *PS'*. The comet fragments and the planet Jupiter are part of the solar system, and in some problems about the career of the fragments at least some of the boundary conditions established by the solar system as a whole must be taken into consideration. We must, for instance, know how many planets there are. Similarly, the total brain establishes boundary conditions for the amygdala, the globus pallidus, or the speech centers of the cortex. Just what boundary conditions need enter into the solution of a particular problem is a relative matter. For most purposes, prediction of the time of impact of two large masses, for instance, need not take into account their electromagnetic properties. But this relativity is a function of our interests: some features of nature may have a negligible relevance to what happens to interest us but may be of great importance in other circumstances. Any (abstract, mathematical) *PS'* we define for precise investigation of our (concrete) *PS* must of course take all this into account.

The relation between an enframing physical system and an enframed one may be summed up in this way: whatever laws of nature prevail in an enframing *PS* prevail also in an enframed *PS*; whatever the parameters of an enframing *PS* may be, those parameters will bear in some causally significant way on the enframed *PS*, even if the nature of the problem under investigation does not require some of the parameters to be taken as boundary conditions of the problem. There is an additional complication that is of some importance: we have a very incomplete knowledge of the laws of nature and of the most remote boundary conditions established by an enframing *PS* greater than any now available to us by empirical means. Indeed, it is not clear that unknown boundary conditions and unknown laws do not in some circumstances amount to the same thing.

A More Commonsense Expression of the Received Doctrine: Identifying Particular Causes and Particular Effects

Laplace's approach to the causality question is far from our commonsense understanding of the relation between cause and effect. If the furnace stops running, we may grudgingly concede that the whole state of the universe immediately prior to its stopping is the cause of the state of the universe that includes that stopping. But we are more immediately concerned to know what *particular* thing or event caused that inconvenient state of affairs. Was it just a touchy circuit-breaker? Or did the oil supplier fail to fill the tank at the scheduled time? We must know because we want to get the furnace running again. Speaking more generally, our commonsense grasp of the notion of cause and effect is closely tied to the notion that human agency can make a difference— that we can cause things to happen in our favor; that we can cause things not to happen if their happening would endanger or inconvenience us. If we have Laplacean views, we set them aside in favor of taking some action that seems to make a difference in our world.

This commonsense understanding of cause and effect is of great importance for science as well. No experiment can be set up without exact attention to the particular cause-and-effect relationships within an instrument or between an instrument and the environment. The ordinary conversation of scientists, not to speak of their ways of explaining their work to the public, relies in many ways on this commonsense understanding of cause and effect. Though well aware that the comet fragments that struck Jupiter formed a physical system, and that this *PS* formed part of a larger *PS* called the solar system, scientists in their public discussions spoke of the fragments as *producing* by their impacts immense disturbances on the planet's gaseous surface. Similar language is commonplace in public discussions of experiments with particle accelerators: it is taken for granted that the accelerated particle will *produce* a spray of short-lived particles when it collides with the particle it is aimed at. Indeed, it is fair to say that the immense and highly complex instrument that is the accelerator is constructed in terms of just such a commonsense image of causality.

Many other examples of commonsense talk about causality can easily be supplied from biology: it is often said, for instance, that the firing of a neuron causes some effect, excitatory or inhibitive, on another neuron with which it is in synapse. Sometimes the singling out of a particular event, condition, or entity as a cause is based on practical considerations: in medical biophysics, for instance, it may be of the first importance to identify a particular defective gene that is a significant causal factor in some disease or physical trait. Yet even in biology the physical systems model defined by Laplace may be a more comprehensive one than the model that picks out a causal relation between particular entities or events. A gene is affected by the particular location in the developing embryo of the cell that carries it: its operative power may be "turned off" or ignored in one part of an embryo while it is accepted in another. To put the matter another way, the enframing *PS* that is the developing embryo sets different boundary conditions for a gene that is an enframed *PS* at different places and times in the course of the enframing system's development. If precise mathematical treatment of what is happening in such an enframing physical system as an embryo is required, the causal model of two sequential states of the same physical system seems more appropriate than the model of two particular entities/events in a $C \rightarrow E$ transaction. The geneticist may not require a complete mathematical analysis and so for some purposes may prefer a model of causality that links two particular entities/events as cause and effect.

In the following informal model for this view of causality, C and E stand for *particular* events, conditions, or entities that are causes and effects respectively, and L stands for whatever law or laws are thought to govern the causal transaction between C and E.

$$L$$

$$C \rightarrow E$$

Philosophers who prefer to define causality in this way sometimes say that the first event, condition, or entity is the cause of the second only if it is necessary and sufficient for the second.[1] In such transactions many things may function as necessary conditions: if a match is to be struck successfully the match must be dry, the striking surface dry (in the case of safety matches the striking surface must also be of a certain chemical composition), and enough force must be exerted to produce adequate friction. If all the necessary conditions are present, they are sufficient to produce a flame. But in such cases the notion of sufficiency seems to be the more basic one, for a sufficient cause will comprise whatever conditions are necessary for it to do its work.[2]

The definition of causality as a $C \rightarrow E$ transaction between particular events, conditions, or entities is older than the doctrine eventually developed by

Laplace. It is in fact very close to the definition of causality that Hume first analyzes and then criticizes. But those who do not accept Hume's criticism may nonetheless consider the definition of causality as a transaction between particulars to be a pragmatic way of operating under a generally Laplacean outlook. Seen in that light, the definition would seem to be compatible with the rationalist confidence about the powers of the human mind that Laplace himself displayed.

It is, however, a historical fact that Hume's empiricist examination of the powers of the human mind—an examination that is the most potent factor in what I have been calling the negative philosophical judgment about the powers of the human mind—was largely based on his criticism of our power to know a causal connection understood in terms of the definition I have just been considering: a connection between particulars (Hume 1748/1955). That criticism repudiates the view that when a particular C is a necessary and sufficient cause, some power in it *necessarily produces* the E, and that, moreover, we can *know* that power. According to Hume, a given C is necessary and sufficient for a given E if and only if the two are always correlated—that is, if E always follows when C happens and does not follow when C does not happen. But there is no *knowable* connection between the two that justifies our saying that a cause is a necessitating power that produces the effect.

But I postpone further consideration of the negative philosophical judgment of the powers of the mind and return to the notion of law that is common to Laplace's doctrine and the $C{\rightarrow}E$ doctrine we have just been considering.

The Status of the Laws of Nature in the Received Doctrine

Twentieth-century philosophers who wish to give a precise sense to the sufficiency of the cause for its effect—either in a setting close to that of Laplace or in the setting that insists on singling out particular causes—usually do so by appealing to the covering/backing laws under which the transition from $PS(s1)$ to $PS(s2)$ (or from particular C to particular E) takes place. The law-governed nature of the transition is tantamount, they say, to the sufficiency of the cause for its effect. The notions of cause and law are thus made interdependent. David Bohm, who has important credentials both as a scientist and as a philosopher of science, expresses the matter in this way:

> In the processes by which one thing comes out of others, the constancy of certain relationships inside a variety of transformations and changes is no coincidence. Rather, we interpret this constancy as signifying that such relationships are *necessary*, in the sense that they could not be otherwise, because they are inherent and essential aspects of what things are. The necessary

relationships between objects, events, conditions, or other things are then
termed causal laws. (Bohm 1957, 1–2)

Sometimes the notion of law is made so important in science that the causal relation is virtually absorbed in it. In an early essay Bertrand Russell went so far as to
claim that the word 'cause' is not needed in the vocabulary of exact science—that
science is not concerned with causes at all but only with laws (Russell 1912).

Many have agreed with him, and some have gone further. In the most
extreme cases of law-worship, time itself—and thus the causal nexus itself—is
considered to be but one more feature of a static structure laid down by transcendent Law. In that setting, causality loses the sense of real temporal becoming we usually associate with it, for time is interpreted as a metric property of a
static structure of which it is but one of the coordinates (Grünbaum 1967). The
nontemporality of the laws of nature is thus understood to be prescriptive in a
causally determinative sense, just as it was for Laplace, rather than merely
descriptive. All other causality is understood to be apparent and in that sense a
"consequence" of the eternal causality of law. The explanatory power of this
model of scientific causality is inseparable from prediction: what is thus
explained could in principle have been predicted. This is especially so in the
Laplacean version, in which cause and effect are simply successive states of one
physical system. Moreover, since the state of the PS called the cause is in turn the
effect of an earlier state of the PS, retrodiction is also thought to be possible in
principle. Modern cosmology, perhaps the most hubristic of all disciplines, is
based on that possibility. There is as much determinism in all this as the empirical adequacy of the statements about the two states of a physical system permits.

Of metaphysical determinism there ought to be none whatever for any
empiricist who remembers that this interpretation of the scientific doctrine of
causality is a practical evasion of an epistemological-cum-metaphysical impasse about necessity—an impasse I consider in more detail later, when we
turn to the second attitude towards the power of the mind. When it comes to
determinism, however, scientists are often metaphysical enthusiasts. I once
heard a distinguished publicist for science, transcending all the practical limitations of science by an imaginative leap, inform his television audience
solemnly that in the first microseconds of the universe the very words he was
then addressing to them had been inexorably laid down. The assumption is
that the universe as a whole in those first microseconds is a (concrete) PS, and
that if the relevant parameters of that state of the PS together with the laws of
nature were known exhaustively, the state of that same PS at the moment of his
talk—including his uttering of just those words—would also be known. Interpretations of the predictive/retrodictive ideal of modern cosmology that

attribute to the laws of nature an ontological status determinative of the very structure of temporality/becoming lend plausibility to such claims.

The resemblance of determinisms of this kind to the determinism Laplace propounded is clear enough. There is, however, an important difference between Laplace's determinism and those of the twentieth century: all modern determinisms must make some accommodation with the fact that, in the physics of the very small, regularities can be expressed only in terms of probability, and that this may be a condition that no amount of additional knowledge about nature can overcome. Laplace, as it happens, wrote extensively and brilliantly about probability and thought it to be of great importance for science: recall that his 1814 exposition of determinism occurs in a philosophical essay on probability theory. But he seems to have thought that the true laws of nature as grasped by the "vast intelligence" towards which human beings move as towards a limit are not probabilistic. For Laplace the concept of chance is defined in terms of our ignorance, and probability theory is only necessary because of that ignorance.

Some modern determinists have argued that the impossibility of giving a deterministic account of the behavior of such entities as single electrons or single photons is of no significance, because the relevant statistical laws are in fact deterministic for the macroscopic level in the sense that from the state at time t_0 of a physical system comprising large numbers of such entities the state of the same system at t_1 can be calculated as rigorously as a simple mechanical state. Others have argued that the behavior of individual elementary particles is indeed deterministic, although the interference of our measuring techniques with what is measured, or other hidden variables in the physical situation, or both, make it appear otherwise.

With a different emphasis, Bohm (1957, 17–64) has argued that the distinction between statistical laws and deterministic laws is a relative one: what in one situation requires statistical analysis will require causally deterministic analysis in a different situation—and conversely. As he says, "the categories of necessary causal connection and chance contingencies are seen to represent two sides of all processes" (29). Bohm's argument is a subtle one, but in my judgment it is a subtle version of Laplace's claim that, whatever our success in science may be, much of what the laws of nature really are remains hidden from us. Perhaps there is no need to pursue this question further at this point; I do take it up again a little later. Just here I am only interested in making the point that the general spirit of the scientific doctrine of causality for much of this century has been deterministic, and that the notion of prescriptive laws of nature has been an important factor in that doctrine.[3]

The Received Doctrine and the Negative Philosophical Judgment about the Power of the Human Mind

The contribution of philosophical empiricism to the scientific doctrine of causality raises a profound difficulty for this deification of prescriptively determinative law and indeed for the whole of the scientific doctrine of causality. Recall that empiricism culminates in this negative philosophical judgment about the functions of the human mind: even if there are in nature beings that have the power to affect other beings causally, the human mind cannot know either those beings as they are or the causality they exercise. Although the scientific doctrine (in the Laplacean version) does not focus so much on the interaction of beings as on the law-governed succession of the states of a physical system, these too would be cognitively inaccessible. It follows from this judgment that if there are indeed laws of nature having a causal significance, the human mind cannot know either *that* they exist or *what* they are. Philosophers of the empirical tradition agree with Hume that there is no justification for the claim that we as knowers can find in nature either necessary production or the lawful necessity of a succession of events. They contend that if necessity does indeed exist in a nature understood to be independent of any formative/constitutive power the knower may conceivably possess, then the knower cannot observe, intuit, or otherwise confront it.

Nevertheless, since philosophical empiricists agree with practicing scientists that the notion of the laws of nature is and ought to be central to the practice of science, they have labored mightily to find a practical equivalent for the necessity that cannot, as they think, really be found in an independent nature. One familiar practical equivalent—the dominant one in philosophy of science—interprets the model of causality given above in linguistic terms and in terms of a logical rather than physical relation: *statements* about the state of a physical system that is regarded as the cause *logically necessitate* statements about the state regarded as the effect.

To put the matter in terms of a distinction I have already made, the topic of causality is displaced from a *PS* setting to a *PS'* setting. This interpretation assumes not only that a body of theory (including laws) is expressed in language but also that the body of theory is part of the very texture or network of scientific language. Statements about a $PS(s1)$ and a $PS(s2)$ are therefore regarded as instantiations of the (linguistic) body of theory, which is to say that they are regarded as statements forming part of a *PS'*. The connection between such statements and the two—presumptively real—states of a (concrete) physical system, $PS(s1)$ and $PS(s2)$, must then be brought about by virtue of semantic rules—rules establishing the truth conditions for the statements. Such rules, however, are understood to be intralinguistic ones, and the entities that are assumed to be ingredients in any physical system—electrons, for instance—are accordingly regarded as things postulated from an intralinguistic situation. This is often expressed by saying that the entities in question are real for that language—alternatively, that they belong to the ontology of that language.

In sum, philosophers in the grip of the negative judgment about the powers of mind sometimes take it for granted that any physical system that can be rationally considered must be, in fact, a PS' (Suppe 1974, 224). Indeed, they take it for granted that any (concrete) PS must have the status of a postulation made on the basis of sensory stimuli/phenomena and our scientific use of one or another PS'. Philosophers who accept this linguistic surrogate for an unattainable causal necessity intrinsic to nature do so at a terrible price: they immure themselves in a linguistic prison of their own making. Although it has no real walls, it is as hard to escape from as a real maximum security facility.

The two attitudes towards mind that have presided over the development of the scientific doctrine of causality clearly differ about the ontological status of the laws of nature. For the first attitude—the confident one—the laws are truly there in the nature of things, and we as knowers are capable of knowing them as they are, although they are not easy to know. The laws have a systematic unity, and we are also capable of finding—although only after much hard work—the underlying unity of what at first seem diverse and unconnected laws. The laws, moreover, are prescriptive (in a deterministic sense): they do not merely sum up the ways of physical becoming but rather mandate, as principles of Being, all the particularities and sinuosities we find in Becoming.

For the second attitude—the one that culminates in a negative philosophical judgment about the reality-attaining powers of mind—the notion that the laws of nature have a systematic unity can only be a hypothesis, and the laws we formulate can only be our own—our own in the sense that we frame them within the linguistic/conceptual situation that is our lot as knowers, and frame them in response to an experience of becoming that itself is qualified by that same ineluctable linguistic/conceptual situation. In view of that origin, the laws must be descriptive rather than prescriptive. They describe, insofar as descriptions formed by a mind about which we have come to a negative philosophical judgment *can* describe, how transitions from one state to another of a physical system—large or small—do in fact take place. The two attitudes disagree, then, on the question of the *ontological status of the laws of nature*. It is a most important question, and I return to it later.

It is, however, curious that the effort in the course of the present century to unify science—to reduce the many laws of nature to a set of laws so simple and elegant that nothing in nature lies outside its scope—has been made largely by philosophers who profess to be in the grip of the second attitude. The old-fashioned reductionist goal for the unity of science, which was pursued zealously for much of this century but is now much criticized, required that all interim laws designed to explain some upper level in a hierarchy of complexity should eventually be deducible from the laws proper to the base level (generally identified with the level of elementary particles) and that all concepts (or terms) applicable to an upper level—such concepts as 'cell' and 'gene', for instance—should eventually be definable without remainder in terms of concepts (or terms) appropriate to the base level. The reductionist unity of science is not a strange goal for philosophers who have confidence in the ability of the mind to get at what is truly the case, and who think that something very like a Laplacean determinism is the case. But it is a strange goal for analytic-linguistic philosophers who profess to dismiss metaphysics.

How can we account for this oddity? I take the risk of suggesting that many of those who profess the second attitude—that is, the diffident attitude towards the powers of the mind—live different lives as believers from the lives they lead as practitioners of science or philosophy of science, and that, as believers, they are as much dominated by the image of a total (concrete) physical system in continuous progression from state to state under eternal laws that mandate just that progress and no other as are those who profess the first, or confident, attitude. The two attitudes, I suggest, coexist not just in the intellectual community at large but often in the same person. That is why so many analytic-linguistic philosophers who embrace an epistemology that forbids any single metaphysical commitment—that is indeed often dominated by the notion that any version of reality must be in some profound sense a linguistic construction—will not even consider any causal account of mind or anything else that is at odds with the scientific doctrine of causality I have been examining.

Summary of the Received Scientific Doctrine of Causality

Despite all the difficulties we have been considering, the received scientific doctrine persists as the dominant doctrine of causality in this century. I summarize it in the following principles.

1. The procedures of science are capable of discovering the causes of things and thus of providing the only adequate explanation of things.

2. The causal relation manifests itself in sequential physical processes. Rigorously speaking, such processes may be regarded as the passage of a physical system from one state to another, the prior state being regarded as the cause, the later one as the effect. Pragmatically speaking, particular events, entities, and conditions may be singled out within these states and regarded as the causes of other and later events, entities, and conditions.

3. Sequential physical processes are governed by laws of nature, and these laws explain why the processes are as they are—indeed, the lawfulness of the processes is what makes them *causal* sequences.

4. The laws of nature can in general be given mathematical form. It is desirable that they should tend towards the Laplacean deterministic ideal, although in practice probabilistic reasoning remains essential in many fields today—thermodynamics, quantum theory, and genetics, for example—and will probably remain so in the future.

5. The laws of nature express the causes of things best when they are unified: if the laws of two different physical domains (for instance, gravitational phenomena and electromagnetic phenomena) can be clearly stated but cannot be brought under another law or laws that reveal a unity underlying their differences, then the explanatory situation is to some degree unsatisfactory.

6. Causality has no telic feature, although human beings tend to attribute teleology (final causality) to nature.

7. Causality has no form-creating feature other than that expressed in the laws of nature.

8. Agency as such, whether divine or human, forms no part of a scientific causal explanation: although human agency is often appealed to in commonsense causal explanations, agency itself is subject to causal explanation in the sense of the preceding principles. Agency is thus something to be explained, not something that is in itself explanatory.

4
Mind and the Scientific Doctrine of Causality

Problems about Applying the Received Doctrine to the Study of the Brain

THE DIFFICULTIES of the scientific doctrine of causality—as understood by those in the grip of the first, or confident, attitude towards the powers of the mind—increase with the complexity of the physical system we try to apply it to. Recall, for instance, that the logic of our model of the scientific doctrine of causality carries with it the notion of nested physical systems, that is, enframing and enframed *PS*s. A human being is an especially complex physical system, which is to say that it enframes many other physical systems, including that *PS* of such daunting complexity, the brain. The *PS* that is the brain in turn enframes many smaller physical systems—associative cortex, visual cortex, thalamus, amygdala, globus pallidus, and so on; but of course even a single neuron in any of these systems is itself a *PS*. The model we are using allows us to think of any of these physical systems as moving from state to state under the laws of nature, that is, under an *L* only part of which is now known to us. Finally, any *PS* we can single out, no matter how large, is enframed by the *PS* we call the universe—a physical system of indeterminate extent. If we continue to follow the deterministic logic of the model, we find ourselves back with the imaginative image of Laplace or with that of the television scientist I mentioned earlier. In principle the evolution of the brain in general and the development of any particular brain also fall under this scheme; and the same thing is true of whatever is happening here or there within any particular brain.

All this, however, is very far from practice: although study of the brain may be theoretically consistent with such a model, research is not conducted in terms of the model. In practice, progress in the study of the brain is made because the brain is approached as a complex and more or less stable structure made of substructures, and because what is wanted at the present stage of research is an extremely detailed understanding of the paths taken through the structure by impulses that are electrochemical in nature. One could consider a single neuron in the course of firing as a physical system moving from state to state in accordance with appropriate laws, and one could consider a structure like the amygdala in the same way, but it is really the structure itself as traversed by impulses that are in some sense guided by the structure that is of central importance. We want to know how all that traffic contributes to rational awareness and rational action. And if we suppose that we are confronted with more than a biological problem—if we suppose, for instance, that we cannot really understand why the traffic is what it is unless we receive from the person in question some information about what is going on in that person's mind— then we also want to know how rational awareness and rational action contribute to the impulse traffic that supports them.

That the total brain is also a physical system and that such a *PS* does move from total state to total state seem true enough, but we have small reason for trying to deal with the total physical system as we imagined astrophysicists to deal with the *PS* made up of the comet fragments and Jupiter, or even as a biophysicist might deal with the very small *PS* made up of an ion gate in a neuron together with the ions moving through it. For one thing, we have not the least idea how to supplement the laws of nature as we now have them (the known part of the ideal *L*) in sufficient detail to provide a causal explanation of the progress of any reasonably complex part of the brain (a *PS*) from a state $s1$ to a state $s2$. For another, we have no way of establishing parameters (state variables) of any such $s1$. In short, the $C{\to}E$ (transeunt) model, in which both cause and effect are regarded as particular events, conditions, or entities, suggests itself: a neuron fires in one location and eventually affects (by way of a stable structure) another neuron located elsewhere.

Brain physiologists clearly assume that the known part of the laws of nature (*L*) that prevail in a physical system that enframes a human body will also prevail in the brain and its substructures. But they also assume that the structure itself *guides* the physical processes that are covered by the known part of the *L*— guides them in somewhat the way a wire or optical fiber guides an electromagnetic impulse. It is important to the investigator, for instance, to know that neural impulses pass from the optic nerves by way of the right and left geniculate nuclei of the thalamus (some having crossed over at the optic chiasma) to the two sides of the primary visual cortex; that some of those impulses then pass on to the associative area of the cortex and thence on down to the amygdala, while others go on to the visual area of the corpus striatum and then on by way of the globus pallidus back to the ventral nucleus of the thalamus. It appears that we can interpret this guidance in at least two ways:

- we can say that these more or less stable structures furnish the boundary conditions under which electrochemical processes take place in accordance with the known part of *L;*
- we can say that the very existence of the structures is equivalent to a new *morphological* law of which my secondhand account of the traffic from the optic nerve is a rough and partial sketch.

Either interpretation leaves us still with the real problem: although there are many tests that assure us that this complex "guidance" of electrochemical impulses is vital to seeing, and although in that limited sense we know some of the significance or meaning of the structures I have listed, we still do not know just how such impulses contribute to our seeing, let alone how they contribute to our being rationally aware that we are seeing something. Moreover, we are left with a problem on which the last word has surely not been said: just how the elaborate structures I have mentioned came into being in both a phylogenetic and an ontogenetic sense.

Neo-Darwinian Evolutionary Theory and the Exclusion of Telic Causality

Consider now just one of these problems, the phylogenetic one. The standard scientific solution to the problem today is that of neo-Darwinian evolutionary theory, which, like any scientific enterprise, must rely on the scientific doctrine of causality to provide its explanatory framework. As we saw, that doctrine excludes both telic causality and any mode of causality that is directly productive of macroscopic forms in a way that is analogous to formal causality. (I argued earlier that the notions of telos and form are closely connected, so to simplify the exposition here I confine myself to telic causality.) In mechanics and electromagnetic theory, the exclusion of the telic by the scientific doctrine of causality tends to be persuasive, because such studies are not primarily concerned with organisms and so do not have to contend with the telic impression organisms make on creatures like ourselves, who are so conscious of the role that purpose plays in our own lives. Neo-Darwinian theorists, however, must directly engage and destroy this inclination to the telic if they are to be persuasive.

Their most important weapon in this engagement is the notion of chance. Statistical reasoning is therefore central to modern evolutionary theory, and it has been extensively and expertly exploited, more often than not with the explicit purpose of dismissing any telic interpretation of evolution. It is argued that, given sufficient time and minuscule cumulative changes that happen "by chance," only the old Darwinian principle of natural selection is needed to account for what might otherwise appear to be a telic process.

If the neo-Darwinian argument were based on a well-founded claim that *absolute* chance occasionally operates in nature, this simple argument for the exclusion of the telic would succeed. But that exclusion would have been bought at great expense to the scientific enterprise as a whole, for the scientific doctrine of causality provides for no absolute chance. To introduce it thus on an ad hoc basis would be to render the doctrine incoherent. The scientific doctrine of causality, whether expressed by those who manifest the first attitude towards mind (the confident one) or by those who manifest the second (the diffident one), is committed only to what I have been calling relative, or Laplacean, chance.

It is easy enough to find senses of relative chance that do not violate the logic of the scientific doctrine of causality. I begin with a simple and obvious case: the chance that you may throw a six with a die is one in six, provided that certain conditions are met. The first condition is that there must be no significant physical bias built into the die. Some bias there will always be: even the depressions sometimes found in an actual die—six depressions on one of the sides, only one on another—constitute a physical bias, but if the thrower is honest, such things are usually negligible. Another condition is the exclusion of any knowledge of the state variables of the physical system that consists of the die, the table, and the throwing hand. Allow that knowledge, eliminate physical bias, and you have an extraordinarily difficult problem in mechanics rather than a case of radical chance. Chance in the die-throwing situation or in any number of situations in the biological and social sciences is clearly relative (Laplacean) chance: it is chance that can be eliminated by better knowledge, although such knowledge is sometimes difficult to come by and hardly worth the trouble.

A more complex and debatable case is that of quantum theory, for in the physics of the very small the elimination of probabilistic reasoning may be impossible in principle. It is true, as I noticed earlier, that some physicists still believe, as Einstein did, that chance of that kind can be eliminated—that a deeper science will find hidden variables that allow, say, a deterministic treatment of the course of single electrons; or that the need to use probability reasoning is a function of the radical interference of the scientist's measuring instruments, and so does not point to a genuine absence of cause/reason at a certain size level. Such interpretations eliminate absolute chance decisively, but there is no present way of deciding whether they are sound.

Other scientists and philosophers of science take the need for statistical methods at the quantum level to be a permanent condition, something that in principle cannot be eliminated. But on the other hand most of those who hold that view do not take this permanent condition to be a symptom of the presence of absolute chance in nature. Nor does Bohm's attempt to do justice to both deterministic and statistical laws, which I mentioned earlier, countenance absolute chance (Bohm 1957, 17–64). In any event, scientists insist again and again that, setting aside the impossibility of following the course of a single electron in certain experimental situations, there is no part of science more precise than quantum theory, no part of science less symptomatic of absolute chance. No doubt the matter is still open, but quantum theory does not seem to support the presence of absolute chance in the universe. Quantum theory may be an odd exception to Laplace's general rule: in this case he may be wrong about the temporary, *faute de mieux* character of statistical reasoning but remain right about the absence of absolute chance.

I began this discussion of relative chance after remarking that if neo-Darwinian theory appealed to absolute chance, and if there were indeed such chance, it could legitimately exclude the telic. But neo-Darwinian theory makes no such appeal. Chance events, or random events, are so called because the laws that express the rate of their occurrence are statistical, and not because they are thought to appear without a causal connection with other events. An examination of the work of Ronald Fisher, the founder of the use of statistics in neo-Darwinian theory, will bear this out (Fisher 1930). G. G. Simpson, who is a reasonably orthodox neo-Darwinian, makes this telling remark, which might well have been made by Laplace: "The cause of an evolutionary event is the *total* situation preceding it (and this in turn the result of total situations on back to the beginning), so that it is not entirely realistic to attempt designation of separate causal elements within that situation" (1953, 59). And in the widely read *The Blind Watchmaker*, a book for the general reader by a well-known biologist, such terms as 'random' and 'chance' are always used in a relative sense (Dawkins 1986, 306–8). It seems fair to say that despite their extensive use of probabilistic reasoning, neo-Darwinians assume a scientific doctrine of causality in which the principle of necessitation from state to state of any physical system is preserved.

It seems that the whole body of law, which is in part unknown and within which the precise relation between statistical and nonstatistical factors is also unknown, is nevertheless regarded as prescribing a smooth, continuous movement from state to state. It seems that the presence of statistical features in evolution is no more suggestive of some failure of the principle of sufficient reason than is the resort to an actuarial table by an insurance company. No neo-

Darwinian theorist supposes that a mutation is random in a sense that demands interpretation by a subtle modern version of the *clinamen atomorum,* atoms that swerve without cause/reason. Such a body of law (*L*) mandates with equal authority the movement of a particle in a field of force, the fall of a weight from the tower of Pisa, the impact of the comet fragments on Jupiter, the transmission of information by electromagnetic waves, the origin (given certain circumstances that are also necessitated) of complex molecules, including the polymeric ones in which the four bases of DNA are entrained. The *L*, it seems, must even have necessitated the extraordinary "encoding" relation by virtue of which an enframed physical system (the DNA) helps necessitate an enframing physical system (the organism) and is in turn conditioned by that enframing physical system. To put the matter in a nutshell, there appears to be no intelligible sense in which that *L* can be said to have necessitated the movement of a particle in a field of force and yet not necessitated the appearance of a Newton or a Mozart.

So, at least, if we suppose that the body of laws of nature (*L*) is deterministically prescriptive in the sense of necessitating the progress of a physical system, of whatever size, from state to state. Such an *L* seems very like a plan; it seems to have been form-necessitating from all eternity—or, if you prefer, from the beginning of the universe. Thinking about it, you may well feel it is understandable that Laplace should have hit upon the metaphor of a "vast intelligence" from which our own must always remain "infinitely removed."

All these form-necessitating consequences of the scientific doctrine of causality are not considered by neo-Darwinian theorists because they do not see the consequences. What prevents them is a convenient but innocent psychological quirk: for all their insistence that relative rather than absolute chance is at work, they give relative chance all the anti-telic force that absolute chance would indeed have. In effect, they take relative chance as the moral equivalent of absolute chance where the question of planfulness or its absence is the issue.

The Question of the Ontological Status of the Laws of Nature

However useful the notion of prescriptively determinative laws may be in science, it may still fail to characterize the nature of things. As I said many years ago, "what we call the laws of nature are inseparable from the facts as we have them; and when we suppose that these same laws have brought about the facts, it is to be suspected that we do not quite know what we are saying. One may say of an isolated system that it is what it is because it belongs to a wider system in which those laws *prevail,* but we can hardly say of the widest system imaginable that it is what it is 'because of' these same laws. The laws are, on the present view, abstract versions of an existent order, and not in any sense producers of order" (Pols 1963, 248–49). In a later book I developed in some detail those reservations about the ontological status of the laws of nature. The following quotation, which occurs in a chapter entitled "Action, Entities, and the Laws of Nature," sums up the alternative provided: "In contrast with the 'real causality' of action it now seems possible to regard them [the laws of nature] as an abstraction from, and a codification of, the ordering power of entities whose ontological status is perhaps more fundamental" (Pols 1975, 35). The ordering power of such entities/beings, chief among them the human agent, was the theme of that book; in discussing that ordering power I often used the expressions 'ontic power' and 'ontic causality' in that book and later ones.[1]

The approach to the status of the laws of nature I have just sketched was unusual in 1975 and even more so in 1963; it is less so now. Though written from a different perspective, John Dupré's *The Disorder of Things* lends that approach considerable support (Dupré 1993). The claim that the laws of nature are derivative from the ontic power of primary beings will, I suspect, receive more attention in the future.

Mind and Microentity Reductionism

In this century the received scientific doctrine of causality has been closely associated with the theme of reductionism. I have already drawn your attention to the apparent failure of what has probably been the most influential form of reductionism in this century—the one that led to the project of deducing all higher-level laws from those of the base level and defining all higher-level entities in terms of a language appropriate to the base level. That project seems to have been abandoned as perhaps wrong in principle and in any event not feasible. But there is another form of reductionism that seems designed to yield a practical equivalent of the abandoned project: it consists in recognizing causal significance only in the entities that are to be found (or postulated) at the lowest, or base, level of nature. It is there, in those microentities, that causality (in the scientific sense) actually *works*.

If we are to think in terms of physical systems moving from state to state, we must regard any *PS* as an aggregate or togetherness of those entities, and we must not take that macroscopic aggregate so seriously as we do the microentities that make it up. If we are to think of the laws of nature as an *L* regulating, prescribing, covering, or backing state changes, we must regard the *locus operandi* of those laws as being just those base-level entities. The laws do not operate, for instance, on or in organisms as such: they operate rather on or in the microentities that make them up.

Science as we know it began, it is true, with laws that apply to macroscopic entities, and of course they do so apply: a human body is subject to all the laws of mechanics. But the form of reductionism I have in mind explains this physical fact and all others by looking to the microentities that make up the macroentity. If any causality is to be assigned to macroentities, that causality must somehow be derived from that of the microentities. As for mind, although it can scarcely be called a macroentity, it is, according to all twentieth-century materialisms, entirely a function of the brain, which is indeed a macrostructure. There is thus a strong tendency to apply the same reductive principle to mind. In short, if any causality is to be assigned to mind, that causality too must be derived from the causality intrinsic to microentities (Searle 1992; Crick 1995).

On the other hand, the human mind, whatever its proper explanation may be, has a tendency or disposition to take macroentities—or at least macroentities of a certain kind—more seriously than that. It tends, for instance, to suppose that human agents can act responsibly and that when they do so there is some causal/explanatory significance in their doing so. That was the spirit of my earlier discussion of causal hierarchies, a discussion I shall soon be resuming. It is just here that the negative philosophical judgment about the power of the mind comes in to reinforce and justify microentity reduction. Philosophers for whom that judgment is now a matter of settled habit suppose that there is something deceptive about our commonsense tendency to see a causal/explanatory significance in macroentities in general and a causal/explanatory significance in the minds that seem to belong to such macroentities as ourselves.

To put the matter roundly, for those who defend microentity reductionism, *the antireductive disposition of common sense is nothing more than a disposition to take an appearance for a reality.* The negative philosophical judgment about the powers of the mind must place the blame for that failing on what common sense calls mind: it is mind that generates appearance, and it is mind that then takes appearance for reality. The negative philosophical judgment tells us that it is of the very nature of mind to respond to whatever it seeks to know by constructing, constituting, or forming something *else* and then taking that something else for what was its original target. If, for instance, mind focuses (at the commonsense level) on what is in fact a complex of microentities, it does not acknowledge that complex but acknowledges instead the presence of a supposed commonsense entity. If it focuses on what is in fact light waves of a certain wave length, it acknowledges instead the presence of a certain color. There are many accounts of how mind deceives itself in this way. For some decades now the most influential account is the one that makes all alleged realities a function of the formative (constitutive) power of language itself. I have discussed this doctrine in some detail in an earlier book and so need not go into it here (Pols 1992, chaps. 3 and 4).

Just here I want to draw your attention to the extraordinary difficulty the many-sided notion of appearance creates for all twentieth-century materialist accounts of mind. For such accounts, mind itself must *be causally generated* (or otherwise explained) by microentities whose operation must be understood entirely in terms of the received scientific doctrine of causality. Mind must also *generate* the appearance that things are otherwise—that is, that such materialist accounts are not exhaustive. Moreover, mind must also *be* an appearance. Yet that same mind must somehow disentangle itself from such causal dependence and such entanglement with appearances—must rise above all that and speak truth about the way things *really* are. It must, that is, be capable of asserting on appropriate grounds the truth of materialism in general and the materialist doctrine of mind in particular. The number of twentieth-century doctrines of mind that try to overcome, evade, or hide this extraordinary difficulty is considerable.

Mind and Hierarchical Causality

We are now in a position to establish a sharp contrast between causality as it is represented by the scientific doctrine and the kind of causality that must prevail in nature if rational animals like ourselves are in fact causal hierarchies. The supposition that we are causal hierarchies allows us to postulate one negative outcome for any scientific approach to the mind-body question by way of neurophysiology. I express the postulation by way of a thought experiment:

- Assume that at the start of a complex rational act a God-like neurophysiologist is able to establish all the relevant physical initial conditions and boundary conditions of the central nervous system of the person undertaking the action and to do this without in any way interfering with the action itself.

- Assume also that the scientist knows all the laws of nature that are relevant to the central nervous system, both those of basic physics and chemistry and also those (if there are indeed irreducibly different ones) that apply to the special case of the complicated biological structure that is the central nervous system. (Such "laws" may amount to no more in practice than a detailed description of the structure of the central nervous system.)

- Assume also that the scientist knows nothing whatever of what is "on the mind" of the person who is under examination—nothing whatever, that is, of what the person is thinking about, intends to do, and will perhaps try hard to do.

The negative outcome imposed on this thought experiment by the supposition that we are in fact causal hierarchies is this: despite a complete knowledge of the parameters of the central nervous system at the beginning of a rational action, despite an exhaustive knowledge of all relevant laws, the God-like neurophysiologist is quite unable to predict the *merely physical* parameters of the central nervous system at the end of the action, for the physical behavior of the central nervous system is qualified by the mind-functions deployed by the apex being.[2] Alternatively, the scientist is unable to offer an adequate *physical* explanation of the state of the central nervous system at the end of the action. Our supposition has of course included the provision that the agent could not have carried out the action in question without the nontemporal (ontic) causal support of the central nervous system; but that provision does not advance the claim that the neuronal happenings are the entire cause of the functions deployed by the agent.

Difficulties about the Notion of Two Distinct Causal Streams, Mental and Physical

I have been concerned, in this chapter and the previous one, to bring out some of the difficulties of a doctrine of causality concerned exclusively with a succession of physical states/events. I now want to return briefly to a topic that is at least distantly related to what we have been discussing in this chapter.

You will recall that I considered, towards the end of Chapter 2, the development, out of Descartes's dualism, of the notion that the mental and the physical may well be understood in terms of two streams of states/events, the one mental, the other physical. That notion has had a profound influence on certain contemporary philosophers of an analytic-empiricist persuasion who feel that materialism does not do justice to the powers of the mind. Wishing to argue that mind does indeed have a causality of its own, these philosophers nonetheless feel obliged to suppose that mind's causality can be exercised only by way of a stream of mental states/events covered by laws of a distinctively mental kind. They are, in short, trying to assimilate a supposed mental causality to the received scientific doctrine of causality.

To this observer, the obsession with this notion rests on a failure to recognize that mental states/events, proposed as objects of philosophical study and assumed to be the bearers of any causality that could plausibly be regarded as mental, are in fact abstractions from the lives of persons. In all plausible cases of what at least purports to be causally significant mental activity, it is only after someone has acted rationally that you can pick out with any confidence a series of states/events (of whatever kind) and consider their causal role in certain purposive achievements. Consider a few rational actions, some impressive, some merely routine:

- speaking a sentence, whether true or false;
- seeing something to be the case—the whiteness of the snow lying in the garden or the cat on the mat—and knowing it to be the case, whether or not one frames a sentence to that effect;

- writing a poem, painting a picture, or composing a piece of music;
- framing a theory about some natural process or structure and then devising an empirical procedure to test the theory;
- understanding and summarizing some difficult text.

Is there not, in any sequence of states/events (mental or physical or both) proposed as a causal account (under appropriate laws) of one of those achievements, a total failure to illuminate the quite different temporal structure of the achievement? The telic unity of the achievement, the directed unity of even one of its several stages (say, the speaking of one word in the course of speaking a sentence), seems to be quite left out. In short, mind itself is left out. In Part Two, I try to call some of these missing features to your attention.

PART TWO

Attending to Mind Itself

5
Mind on Its Own Functions

Where We Stand

THE SUGGESTION that a rational agent is the apex being of a hierarchy of causes and so a *primary being* remains still to be justified; but from where we now stand the claim does not seem to be in radical conflict with the findings of science. The actual practice of science is in fact compatible with an interpretation of the laws of nature that gives ontological preference to hierarchical causality. The laws of nature may indeed be what—according to the epistemological premises of empirical philosophers as distinct from their attempts to evade the consequences of those premises—they must be: regularities extrapolated to a universality that ranges far beyond their empirical base. So interpreted, the laws of nature are descriptive rather than ontologically determinative; and what they describe is the reiteration of the functional behavior of beings whose causal structure is more concretely and more adequately understood in terms of a hierarchy of causes.

The more numerous the entities forming the base of the description, the more pervasive the laws, a consideration which accounts for the precision (and perhaps also the deterministic character) of the laws of physics when compared with laws based on entities of greater complexity. The same consideration allows us to entertain the possibility that the de facto recourse to statistical laws in many branches of science is a permanent condition of science—something not to be removed by a gradual approach to the Laplacean limit in which only deterministic laws are the rule—and moreover allows us to entertain that possibility without conceding the presence of absolute chance in the universe. There is no need to postulate some subtle modern version of the Epicurean *clinamen atomorum*—atoms that swerve without cause, without reason—to explain why some laws of nature, and perhaps all, are ineluctably statistical. We need only suppose that an apex entity can determine what is determinable in the pyramid in which it expresses its self-identity, and that the more complex that self-identity is, the less predictable the outcome.

This chapter and the one that follows it are attempts to direct the attention of both writer and reader to the concrete actuality of mind itself. Since our theme is embodied mind, we acknowledge the presence of an infrastructure, and we acknowledge as well its massive—indeed its tragic—importance in human life. But in these chapters we are intent on deploying mind itself: we are *using* the infrastructure rather than just studying it. Yet similar acts of mind itself, involving a similar subsumption of the infrastructure, would be necessary if we were engaged in the scientific study of the infrastructure.

In this chapter we address ourselves to the functions of the minds of the apex beings (primary beings) we are. In the next chapter we address ourselves to the unity that expresses itself in these several functions, namely, the causality of the apex being that is refracted in these many functions. In this chapter the *functions* of the being occupy us; in the next the *being* itself occupies us.

We address ourselves first to the mental functions of human beings because a long philosophical tradition alleges that both beings and their causality are beyond our cognitive powers. What amounts to the same thing, the tradition alleges that beings and their causality are appearances generated by what purport to be our cognitive powers.[1] In the course of this book I have said that this tradition culminates in a negative philosophical judgment about the powers of the mind. In this chapter and the next I ask your cooperation in trying to undo that judgment.

What are we doing when we attend to mind itself and try to discriminate its functions? I suggest a tentative answer that only the sweep of both chapters can make good: we are deploying the causality intrinsic to the beings we are in order to remove a doctrinal obstacle to the acknowledgment of that causality. In other words, the business of both chapters is the functions of mind and the unity that gathers them together, uses them, and reflects upon them. The reflexive turn begins here with the functions of the mind, but its telos is an apex being in which the causality of embodied mind is already intrinsic.

I consider the reflexive turn in four movements, but that is an expository device, and it does not conform to the actuality of the reflexive life of our minds. We make the turn in our several different ways. What is important is a persistence in and an intensification of the turn wherever it begins. I place three of the expository movements here in this chapter; the fourth I place in Chapter 6, where it serves to link the theme of the several functions of the mind with the theme of the unity of the primary being that deploys the functions in action.

The Reflexive Turn, First Movement: The Functions Described

When you try to list the functions of the mind you are already in a reflexive mode. Some functions of mind must be brought into play to discriminate and identify the functions; in turn, those discriminating functions must be identified and placed on the list. It is no easy or straightforward task. The resemblance of the task to the one Descartes embarked on in the *Meditations* is no accident, but throughout this book I have argued that a doctrine or theory, even one produced by a mind whose greatness is beyond question, can stand in the way of our engagement with the actuality we are intent on knowing. So it is with Descartes's doctrine: I argued in Chapter 2 that Descartes's list, taken together with his elucidation of it, has had disastrous consequences for our understanding of the human mind. We must, I suggest, redo the enterprise of Descartes. Cézanne once said, of a great predecessor in his own trade, that he would like to redo Poussin, but on the basis of nature: "Je voudrais refaire Poussin, mais sur la nature." I make bold to say, "Je voudrais refaire Descartes, mais sur la nature."

It is obvious that if you produce a reasonably adequate list of the functions of the mind, you must have done so by virtue of the reflexive deployment of at least some of the functions that appear on the list. I have already made a positive suggestion about the causal structure of the functions of the mind: we are causal hierarchies in the sense I am trying to draw to your attention and my own. If that causality is authentic, then any deployment of the functions of the mind is an instance of it. In that sense, the deployment of a function of the mind both depends on and commands that part of the causal hierarchy I have called the infrastructure.

Whatever the functions of the mind may be, we can deploy them without understanding just what role the infrastructure plays in the functions. Moreover, when in reflection we turn some of the functions back towards themselves, the success of the reflexive act in no way depends on our being able to give an adequate account of the contribution of the infrastructure to that act. The functions of our mind are reflexively accessible to our attention even though most of the infrastructure elements that support the act of attention are not. All the conditioning support exercised by billions of neurons—indeed, all infrastructure support—disappears, so to speak, into the reflexive exercise of the function, leaving visible only the function itself and whatever causal authority is intrinsic to completing its mission.

Here, then, is a list of the mind's functions, not perhaps as comprehensive as it could be, but more comprehensive than most such lists: mind knows, makes (that is, forms, produces, creates), understands, thinks, conceives, perceives, remembers, anticipates, believes, doubts, attends, intends, affirms, denies, wills, refuses, imagines, values, judges, and feels. You will notice that this list includes not only everything in the famous list Descartes used to define a thinking being in Meditation II but several other functions as well.

The first item on my list, the function of knowing, was not on Descartes's list, and at first glance it is hard to see why. It is perhaps the most basic of all mind's functions: the one we use most, the one we take for granted, the one that Descartes so took for granted that he failed to notice it, the one that empiricism also fails to notice. There is also one striking omission from my list: in our day one expects to find consciousness, or awareness, on such a list, even though it is not on Descartes's. It is, however, a most important and pervasive feature of mind, so pervasive that it does not fit easily into such a list. In one way or another consciousness, or awareness, qualifies all of the functions I have listed, and at least some of them—knowing, doubting, attending, and feeling, for instance—may fairly be described as modes of consciousness or modes of awareness.

The absence of the term 'consciousness' from the *Meditations* does not mean that Descartes is not concerned with consciousness in that momentous book. Of course he is, and the use of 'consciousness' by Locke and other philosophers later in the seventeenth century is surely due to Descartes's emphasis on the inwardness we tend to call consciousness today.[2] Nevertheless, Descartes does not employ the term *conscientia* in his Latin version of the *Meditations,* and the French term *conscience,* which like the Latin *conscientia* can mean either conscience or consciousness, does not appear in the French version of the *Meditations.* (Descartes approved that version, although he did not make it.) Consciousness is so pervasive a feature of mind that we are tempted to say that the terms 'mind' and 'consciousness' should simply be considered synonyms. On the other hand, consciousness is a feature of mind rather than identical with it, for mind does some things unconsciously, and it can—to some degree at will—bring things into consciousness or dismiss them from it.

This brings us to a function I deliberately omitted from the list: the very basic function we call action. Many and perhaps most actions seem to involve the mind, and some of them draw so heavily on mind that we may well regard them as acts of the mind.

When mind summons something to consciousness or dismisses something from it, mind is clearly acting, or in any event the person is acting by virtue of mind. Often enough, acts of the mind have an overt component: you might, for instance, banish something from consciousness by saying, "Let's change the subject," or, "Enough of that." But sometimes they do not, so we must think of action in a sense broad enough to include much that is not visible or audible. Mozart was clearly acting in the example I gave in the Introduction. But his action was not confined to playing solitary billiards: the most important part of the action I have imagined, his composing, took place in his mind and was neither seen nor heard.

Sometimes this or that function on the list will also count as an action. When we make something, whether a theory, a poem, or a piece of furniture, we certainly act; and attending, refusing, denying, and judging can all be acts, and sometimes very deliberate ones. But on the other hand, action seems to be a function of the whole entity/being—perhaps the most comprehensive of its functions. We may well think of action as expressive of the very unity of a rational animal: in the case of rational actions that unity would then pervade all the functions of the mind we are able to discriminate within the action. I pursue this theme further in Chapter 6.

Although the functions we are considering are distinguishable one from another, they are not completely discrete and separable. Mind has a certain unity, and although that unity is difficult to elucidate, one mark of its presence is the way in which various functions overlap. Knowing, for instance, is closely related to understanding and thinking, and perhaps those two functions are best understood as knowing at work in certain special circumstances. Knowing itself is a matter of degree: we may know certain things about a complex situation and be quite ignorant about the rest. Understanding suggests an achieved clarification of what had in some ways been obscure: a kind of advance in knowing in the face of difficulties. As for thinking, it may well be regarded as knowing at work in a systematic and discursive way.

Knowing overlaps other functions by bringing them into play: it always seems to involve anticipation and remembrance, for example. As the activity that generates knowledge, it takes up time, holding fast to what is gone even as it anticipates the future; and it is in part in virtue of that temporal reach that it can grasp temporospatial things. Knowing is also a way of attending, one among many ways, for not all our attention to things is merely cognitive; it involves intending as well: intending to know and to articulate what is thus known. Finally, knowing unites conceiving and perceiving in a way that is so far not clearly understood, so much so that our very abstracting of the functions of conceiving and perceiving from the mental function of knowing does some violence to the unity of knowing.

The theme of action reminds us that this is no Cartesian reflection, even though the list of functions we are considering includes the items that were on Descartes's list. Although we can distinguish the functions of the mind from the neurophysiology that supports it and from other elements of the infrastructure, the fusion of the unity of the function with the billionfold multiplicity of its infrastructure elements is embodiment itself. It is a union so close that the notion of a *relation* between mind and body does not do it complete justice.

This present reflexive examination of the functions of the mind proclaims their embodiment. Now, as I write in an old-fashioned way, with a pen, my hand drives the pen across the page as the thought develops; and the development of the thought, though presumably mental enough, is integral with the driving of the pen. Later, as I turn to a word processor for successive drafts, my hands will drive the machine, but the driving of it will still be integral with the thought. Indeed, language itself, which is made by mind, has been from its earliest beginnings not only a thing of meanings but a thing of sounds and shapes as well.

Given the profusion of acts of the mind that are intricately and ineluctably embodied, the notion that mind can be adequately described in terms of sequences of purely mental events set in contrast with sequences of purely physical events taking place in the brain seems an unreal contrivance. The contrivance is based on the obsessive notion that any physical event singled out from a physical system is wholly caused by prior physical events. Take that notion and apply it to supposed mental events and you have a philosopher's abstraction leading to a straw-man dualism that can then be easily discredited in favor of one of the many forms of physicalist monism that are current today. (See the last section of Chapter 4.)

The Reflexive Turn, Second Movement: Two Key Functions of the Mind

In this movement of the reflexive turn, reflexivity is more self-consciously brought into play, and its originative character begins to be recognized and systematically exploited. The reflexiveness of mind does not inevitably mirror some previous achievement of mind. At least in the case of mind's most vital function, the reflexiveness of the function leads and intensifies it, bringing out possibilities in it that were hidden by a routine or habitual employment of it.

That most vital function is the one that stood first on my list, knowing; in this phase we focus our reflexive attention on that function and on the one that stood second, making. Let me now call these two functions *direct knowing* and the *formative function* respectively. It is important to notice, however, that in this movement of the reflexive turn the function of direct knowing will be the one that is chiefly in play. Indeed, if reflexiveness is a natural and familiar capacity of the human mind, it is so precisely because it is a capacity of direct knowing.

This movement, then, is designed to clarify the central functions of the mind by way of cultivating and intensifying the reflexive capacity of direct knowing. There is, I suggest, no other way to make mind itself aware of its own prerogatives. In that sense, the formative function, and indeed all the other functions of the mind, can be illuminated only by way of the function of direct knowing.

On the other hand, I am writing a book, and a book is an artifact and so a product of the formative function. An expository order is essential to such an artifact, and so it is wise to give in advance some sense of the goal (the telos) I have in mind in this part of the book. The primary goal, obviously, is to overturn the negative philosophical judgment about the powers of mind by exercising and exhibiting the function of direct knowing. But I also intend to contrast direct knowing with indirect knowing towards the end of this movement, and to do so with some precision.

It may be helpful to anticipate right now two features of that contrast: first, indirect knowing is parasitic on direct knowing; second, indirect knowing takes place by way of certain products of the formative function, so my discussion of indirect knowing will include an account of the formative function as well. Simple examples of direct and indirect knowing are also in order here at the beginning of the movement. I know directly my pen and paper and the contents of this room; you know directly the page you are reading and whatever other commonsense objects, natural or artificial, are present in the place where you read. On the other hand, our knowing of such things as electrons is entirely indirect. We can know *theories* about electrons directly, and we can know macroscopic *evidence* about the presence and nature of electrons directly; but electrons we cannot know in the direct way in which you and I could know each other if we were talking together. Finally, let me anticipate the role the formative function plays in indirect knowing: theories are products of the formative function; language, which is essential to the development of theories, is also a product of the formative function.

Direct knowing is a homely and ubiquitous function that seldom fails of its target in a commonsense setting. Like common sense in general, direct knowing can be in error; but closer attention to the matter at hand is usually enough to rectify its mistakes. Just now we are not operating in a commonsense setting. We are making direct knowing the object of our attention and at the same time *using* that same function. We are deploying the function reflexively to establish its own accomplishments, rights, and limitations. I begin with *primary direct knowing*—that is, the knowing of concrete things that are available to us by way of our senses. I distinguish that kind of knowing from *secondary direct knowing*—that is, the direct knowing of such things as doctrines, theories, concepts, propositions, words, and language in general. Most of the discussion of secondary direct knowing comes later, although I must touch upon it occasionally in drawing your attention to primary direct knowing.

The most important fact about primary direct knowing is that it consists of an intimate union of rationality on the one hand and experience/awareness on the other. Philosophical approaches to the activity of knowing tend to do violence to that intimate union. Philosophers are moved to assign rationality to the sphere of concepts and linguistic propositions and to assign experience/awareness to the sphere of sense perception and feeling. Those steps once taken, it is extraordinarily difficult to bring those spheres together again: philosophers suppose themselves to be *using* concepts and language to *organize*, *form*, or *shape* raw material received through sensation; they suppose themselves to have made or formed the things that naive common sense supposes itself to know directly—the tree in the garden, the cat on the mat. The reflexive turn is intended to overcome this deficiency by attending to the union of rationality and experience/awareness and seeing that union for what it is. I hope to persuade you that it both takes place in the knower and completes itself in the thing known. We do not bring that union into being by the reflexive turn, for it has always been there to be found; but in reflection we overcome our long habit of relying on the union without acknowledging it.

In acknowledging the union and cultivating it we deepen it, and it is only in that restricted sense that direct knowing is originative. I make this point to avert a possible misunderstanding: direct knowing is a realistic function and does not inevitably form, make, produce, or constitute what it knows. It is the formative function that is productive, and although in certain circumstances direct knowing cooperates with the formative function, the missions of the two functions should not be confused.

Notice that we are not at this point concerned with *knowledge* understood as what can be expressed in propositions. We are concerned rather with the *activity* of direct knowing and with what that activity completes itself in. Although the activity may be accompanied by the production of propositions, primary direct knowing completes itself in concrete actualities rather than propositions. I hope to persuade you that all propositional knowledge—hence all bodies of doctrine/theory—depends on the activity of direct knowing (both primary and secondary). To say so is to claim that the formative function of mind, which produces bodies of doctrine/theory, operates only on the basis of things known directly (instruments, readings on instruments, symbols on a page), not on the basis of things merely "experienced" (atomic sense impressions or stimuli). The relations between the formative function and the function of direct knowing in the producing, sustaining, and authenticating of bodies of doctrine/theory are exceedingly complex. I return to them somewhat later.[3]

Whether primary direct knowing focuses on the papers and other things on my desk, on the cat drowsing on a small oriental mat before my fire, on what I saw a moment ago from my high study windows—a plantation of gnarled old lilac bushes dark against the first snowfall in late November—or on whatever you now have before you in a different time and place, the function comes immediately under its own reflexive glance even while it focuses on such things. Begin, then, with concrete things that are immediately present to you, but do not do violence to the integrity of the function you are now deploying: let your reflexivity attend to that integrity even as you attend to whatever is before you. Above all, do not yet try to analyze the integrity of direct knowing.

Do not, for instance, try to break it up into the discrete components that so obsess epistemological theorists. The rational component is typically identified as a selection of such items as concepts, ideas, categories, representations, terms, and propositions. The empirical component is usually identified as sensory experience, and it is typically categorized as impressions, sensations, perceptions, feelings, and the like. Such cognitive intermediaries are alleged to be found within mind (regarded as consciousness/subjectivity) and to exist there, in the "rational" case, as things known or, in the "empirical" case, as things merely experienced. It is then claimed that the process of knowing concrete things like lilacs, cats, firewood, and books only begins with such in-the-mind things, the concrete things being indirectly known (inferred) by mind, believed in by mind, postulated by mind, or else constructed within mind by some productive or object-constituting activity of mind. This line of thought develops apace with the line of thought that leads to what I have been calling the negative philosophical judgment about the powers of the mind.

A sound reason for not falling into this conventional professional trap is the impossibility of carrying out that kind of analysis without immediately reinstating the seamless unity of the function I am calling your reflexive attention to. In your analytic effort you would be deploying your power of direct knowing on the one hand towards a "rational" item like a concept or category and on the other towards an "empirical" item like a particular patch of color or supposed impression of a color. You would be drawing on the unity of the function of direct knowing to *know* both concept/category and color patch/impression. In short, you would be depending upon the unity of the function to deny its unity. And if the function could legitimately be broken into discrete "rational" and "empirical" parts, you would then have the task of analyzing both your knowing a category and your knowing that you are perceiving a particular patch of color into yet other pairs of rational and empirical components.

You cannot suspend the seamless unity of the act of direct knowing when you undertake analysis, anymore than you can suspend the embodiment of mind when you engage in abstract logical reasoning. You can only do what is so often done in theory of knowledge: fail to notice what you are actually doing. The point is that the function of direct knowing operates in all sorts of circumstances and on all sorts of objects. Whenever it is brought into play, it manifests factors that are real enough yet do not exist in a "pure" form. When I know the lilacs (by attending to them) I am also actualizing, among other functions, the functions of conceiving and perceiving, but I am actualizing them together, as part of the integrity of (primary) direct knowing. Experiencing (in the restricted sense of perceiving) does not vanish when (in secondary direct knowing) we attend to rational items like ideas and concepts. Rationality (in the restricted sense of conceiving) does not vanish when (in primary rational awareness) we attend to experiential items like patches of color.

One important use of language is to call our attention to something that is accessible to us and is not itself linguistic. But we also use language in the formation of theories, postulates, and hypotheses—all in some sense linguistic—about things that are not accessible to us, and we are therefore inclined to suppose that all knowing must be indirect and theoretical. When this happens a change in terminology may help us reassert the attention-directing function of language. With that in view, let me now substitute the term 'experiential' for the more usual 'empirical'. I can then insist that your reflexive attention to the integrity of knowing will show the union of its rational and experiential aspects to be *sui generis*.

For human beings the experiential always suffuses the rational and the rational always suffuses the experiential: we do not know the lilacs by knowing the word 'lilac' (or the concept of lilac) on the one hand, by experiencing some sensations (perceptions, impressions) such as color patches on the other, and then putting these supposedly discrete rational and empirical findings together. If we try to separate rationality and experience by analyzing what is going on in terms of conceiving on the one hand and perceiving on the other, we overlook this mutual support.

Our failure, however, can be made to leap out at us. If, for instance, we focus (in secondary direct knowing) on a concept or proposition we suppose to be an indispensable mediator of knowing, we soon find that we do so by virtue of rational-experiential engagement with the item we have focused on. We are not, that is to say, deploying a pure, formal, and nonexperiential rationality towards concept or proposition as a preliminary to categorizing experience.

Our very entertainment of concept or proposition is thoroughly pervaded by the experiential—indeed, the embodied—character of all human awareness. There is thus an experiential component in our (secondary) direct knowing of the concept no less than the conceptual component that we set out to isolate.

If, on the other hand, we focus reflexively on ourselves in an act of what we take to be perceiving—say, our attention to the distinctive color of a lilac bud as it is about to come into bloom—we find that we are *rationally* aware of the color as such and also rationally aware of our perceiving it. There is thus a conceptual no less than an experiential component in our awareness of the color.

The change in terminology from 'empirical' to 'experiential' provides us with an alternative name for direct knowing: *rational-experiential engagement*— engagement, that is, with the real, or being. But the most vivid name I have found for direct knowing is the one implicit in the previous paragraph: *rational awareness*. That expression reminds us that attention to something on the part of a conscious subject is central to knowing, and that the subject's attention does not necessarily oscillate between an "empirical" or sensory presence of that particular something and some peculiarly "rational" general item, such as a concept. The subject's attention can focus instead on the integral union of the two factors in that one thing. As we shall see, that integral union of the two factors in the thing known is answered by an equally integral union of the two factors in the knower.

A Useful Terminological Digression: 'Subject', 'Subjectivity'

There is a sense in which consciousness, subjectivity, and awareness are roughly equivalent, and once that sense is established we shall have two more useful names for direct knowing. The term 'subjectivity', however, often brings negative overtones with it today. A terminological digression may help suppress those overtones. The point of the digression is not simply to rescue a term. The point is to rescue from oblivion a positive and concrete function of subjectivity that has been dismissed from serious consideration in great part by the misuse of the term.

The development of the notion of subjectivity is interwoven in the history of philosophy with the question whether what has been called by such names as the 'self', 'ego', 'subject', 'soul', and 'person' really exists as a being independent of the knower or is in fact a postulation, projection, or construction that owes its apparent reality to some formative power intrinsic to the mind of the knower. More often than not, philosophers have concluded that if mind supposes there really are in nature beings appropriately called selves—beings characterized by continuity and by a unity that controls their development and manifests itself in their functions—then mind is mistaken.

You will recall that philosophers who make that negative judgment about mind tend to use the term 'substance' rather than 'being': they do not argue that the self (ego, subject, person) is not really a *being* existing independently in nature and known as such by the mind but rather that it is not a *substance* having that status. They do so because 'substance' has been the traditional translation of Aristotle's term *ousia* since shortly after the beginning of the Christian era. Aristotle's *ousia*, however, is a noun formed from the feminine participle of the verb 'to be', as I have already pointed out, and it is quite accurately rendered in English as 'a being'; indeed, some recent translators have adopted that translation with happy results. There are many good reasons for going back to 'a being'. Chief of them is that 'substance' suggests a static, inert, and merely material continuity of the self, which is surely the wrong thing to be looking for. The term 'a being' allows us to look for the integrity of a patterned development, the continuity of a directed order, which is certainly what Aristotle had in mind as the nature of a human being—that is, a human *ousia*. (See Chapter 1.)

The term 'substance' is nevertheless of great importance, if only because its history is closely connected with that of the term 'subject'. But the Latin-based term 'subject' is properly used to translate Aristotle's term *hypokeimenon*, which he applied to any concrete being (*ousia*) regarded as possessing properties that were less fundamental than the being itself: the being *lies under* its properties, as the Greek word says, and it persists even though its properties change. A person's skin color, for instance, might change with exposure to the sun; a person who cannot play the violin might over the years become an expert violinist. The term 'subject' thus became part of systematized grammar and logic: the concrete being/entity/thing was well suited to be the *subject* of a sentence; its properties well suited to be *predicated* of the subject (Aristotle, *Categories* 2a11–4b19).

It is important, however, to notice just what Aristotle is doing: he is not making *ousia* synonymous with *hypokeimenon*. He is rather saying that one aspect of an *ousia* is that it persists through certain changes and so can be regarded as the subject of the change—that is, as what *underlies* the change. But the notion of unchanging persistence that goes with the term 'subject' (*hypokeimenon*) came to dominate the essentially dynamic, developmental notion of living beings (*ousiai*) that Aristotle presents in *On the Soul* and *Nicomachean Ethics*. No doubt this subject aspect of a being contributes to the fact that in Latin *substantia*, which means 'that which stands under', became the standard translation for *ousia*. Given its meaning, *substantia* is obviously a poor translation for *ousia*, but the meaning of *substantia* is close enough to the meaning of both 'subject' and *hypokeimenon* to be faithful to one aspect of concrete beings.

There are other factors in the early success of the term 'substance' in European philosophy, but they are difficult to illuminate without an excursion into

Trinitarian theology that seems unnecessary here. In any case, successful it was, and as a consequence generations of philosophers have thought that if they do not find an absolute and unchanging continuity in the human being that is the focus of their cognitive attention—if, that is, they have not found a *substance*—they have not found a being that can be called a self, ego, subject, or person.

I noted above that the original meaning of 'subject' does not imply consciousness (later called subjectivity) or even sentience. Although the origins of the notion of subjectivity are to be found in Descartes—more precisely, in the response to Descartes's philosophy by the philosophers who came under his influence—he himself did not invent the term 'subjectivity', and in fact it does not occur until much later. The word *subjectum* occurs only twice in the Latin text of the *Meditations*. One occurrence is irrelevant to the point I wish to make; the other (Meditation III, para. 14) refers to a physical thing—something in which heat may be produced (Descartes 1641/1990, 131). The reference is so clearly physical that J. Cottingham translates *subjectum* there as 'object' (Descartes 1991, 2:28).

Nonetheless, what Descartes calls a thinking substance can just as well be called, on the basis of the medieval terminology Descartes worked with, a thinking *subject*. Descartes grudgingly says as much in replying to an objection made by Hobbes.[4] Thus we can say that for Descartes ideas have their only actual reality as actual features—Descartes often calls them modes—of that subject/substance, the mind. But Descartes clearly prefers, both in that passage and elsewhere, to speak of ideas as modes of the *substance* mind. The important point is that for Descartes a mind has no exclusive right to be called a subject: there are also physical subjects and they too are more fundamental than the modes—heat, for instance—that inhere in them.

The later meaning of 'subjective' and 'subjectivity' is quite different. The change came gradually: the first occurrence in English of 'subject' and 'subjective' in a sense that foreshadows the change is in 1682 and 1707 respectively (*OED*, 2d ed.); the transformation is complete in Kant's work of the 1780s. But clearly the change was in great part generated by certain other things Descartes says about ideas and other "modes of thought." He tells us that the only things mind can know directly are itself and its own ideas. But ideas also have *representative reality* in addition to their actual reality as modes of the actual reality of the subject/substance called the mind. Descartes supposes that we can know things other than mind only indirectly, by way of their representations in the ideas of the mind.

Thus if subjects called bodies really exist—both the bodies that seem to make up the commonsense world and our own particular bodies, considered as parts of that world—we can only know *what* bodies are (their true natures) and *whether* bodies really exist independently of mind by undertaking a demonstration that begins with the representative reality of our own ideas. In the *Meditations* Descartes devotes most of the last and in many ways the most difficult meditation, the sixth, to this demonstration. It is important to notice, however, that what such a demonstration purports to demonstrate is that bodies really do exist as (concrete and actual) *subjects*. (See my discussion of Meditation VI in Chapter 2.)

All this said, we are ready for another terminological complication. Because of the terminology Descartes inherited, the expression 'representative reality' is interchangeable with 'objective reality'; the thought is that what is represented to the mind by one of its ideas is in effect an object *for the mind*. As you can see, then, the way we tend to use the terms 'subjective' and 'objective' today in commonsense talk is in some ways an inversion of the Cartesian sense. Forgetting that for Descartes both minds and bodies are subjects, and attending only to the fact that Descartes's own argument begins in the subject called the mind and only later, and with difficulty, finds its way to the existence of subjects called bodies, philosophers began a train of reasoning that in due course brought about that change in meaning.

Today, if we call a conclusion subjective we mean that it may not in fact express the way things actually are outside the mind of the person who draws the conclusion. We may also take it for granted that the conclusion is based on feeling rather than on sound reasoning. If, on the other hand, we call a conclusion objective we mean that it expresses what is in fact the case—what is real in the sense that it is independent of the mind of the person who draws the conclusion.

It may be helpful to summarize here Descartes's doctrine of the representative character of all knowing, for that doctrine is the most important factor in the development of the dominant modern notion of subjectivity. I summarize it without the late medieval terminology this shaper of the modern era uses in some of the most crucial parts of his work. (See also Chapter 2.)

- The only way to things (beings, realities, existences) is by way of ideas: ideas can be directly present to (directly known by) mind, but realities other than ideas cannot be so present. (Descartes gives the term 'idea' a much wider extension than is usual today: he counts as ideas, for instance, many things that later writers call impressions.)
- Ideas purport to represent things (beings, realities, existences) that are not ideas.
- Some ideas adequately represent things that are not ideas; other ideas fail to do that representative job, either because they are not adequate to

what they purport to represent or because what they purport to represent does not exist at all.

Stated thus without qualification, the representative theory is simply not true. A much more modest claim—that knowledge is sometimes based on representations entertained by the mind, that moreover mind itself plays a decisive role in the making of such representations—is true. Even that more modest claim requires a further qualification: the representations are seldom ideas, or at least seldom merely ideas.

No doubt the truth of this more modest claim is the springboard from which so many philosophers make the fatal leap to the unqualified representative theory of knowledge. Since the unqualified theory first emerged, to dictate in the end the whole course of Western epistemology and metaphysics, it has prevented us from confronting and acknowledging this momentous truth: in mind's most fundamental activity/function (direct knowing, or rational awareness) representation plays no role whatsoever. If representation were inevitably vital to knowing—if, that is, there were no direct knowing—we should be in no position to produce representations and make use of them in circumstances where they are relevant. The role of representations in human affairs is considerable, but we can neither contrive representations nor make use of them if we cannot know some things that are not in any sense representations.

The representations I have been considering are in some cases representations for the knower, and in other cases they are alleged to be such. There are also, of course, representations that are not available to the knower as such but nonetheless play some role in knowing. Neural codings, or mappings, for instance in the visual cortex, are available to the neuroscientist, who can see that they are in some sense isomorphic with something outside the brain; but they are not available *as known or perceived representations* to the human being in whose brain they are found.

What I have given so far is a simplified account of the origin of the dominant modern notion of subjectivity; the history of the terms 'subjectivity' and 'objectivity' is exceedingly complicated. A major complication was introduced by Kant in the late eighteenth century, by which time the change in meaning I have so far described had largely taken place. Kant made a further deliberate change in the relation between the terms, and although this change in emphasis contributed positively to the modern notion of subjectivity—it was indeed a decisive factor in the Romantic movement and in the related development of philosophical idealism—it also introduced a qualified sense of objectivity that makes today's commonsense association of 'real' and 'objective' questionable.

In Kant's view, objectivity—the world of common sense and of Newtonian science as well—is subjective in origin (Kant 1781/1787/1929, 25 [Bxxiii], 147 [A125–26]). Originating thus in the cognitive powers of the human subject, it cannot be real in any ultimate sense. Kant therefore calls this objective world phenomenal: it is, strictly speaking, an appearance rather than reality. In some respects Kant is going right back to Descartes: he even uses the term 'representation' about the objective world of common sense. The German word for representation makes this point especially vivid: it is *Vorstellung*—literally, something *placed before* the knower *by* the knower. For Kant, the very outwardness, or external character, of the ordinary world is thus also something placed before the knower *by* the knower; that is part of what Kant means by saying that objectivity is subjective in origin. Observe, however, that Kant is not talking about *particular* subjectivity: that pattern of feeling, preference, inclination, and so on that distinguishes you from me. In Kant's language, the nonparticular subjectivity all particular subjects share in is *transcendental* subjectivity; it gives objectivity the pattern of universality and necessity that common sense takes to exist independently of the knower (ibid., 337 [A355]).[5]

In other respects, as Kant clearly acknowledges, transcendental subjects remain particular subjects: they have their own patterns of feeling, sentiment, taste, and preference. In short, considered as particulars, they possess the kind of subjectivity we have in mind when we think of the literature of the late eighteenth and early nineteenth centuries as subjective.

Reflexive Turn, Second Movement, Continued

After that terminological digression we may now associate the terms 'consciousness' and 'subjectivity' and use them in a positive sense to provide two more names for direct knowing: *rational consciousness* and *rational subjectivity*. As I use these expressions, they do not call attention to an epistemic predicament of self-enclosure in which the knower is either cut off from reality entirely or else compelled to settle for the (phenomenal) reality that is Kantian objectivity. There is no such inevitable predicament, although in certain circumstances subjectivity can indeed be a predicament: we can entertain a bad theory and be so taken in by it that we are quite unable to see what is really the case, even though what is really the case may be readily available to us. Descartes's representative theory of knowledge, for instance, is not just a theory about representation: precisely as a theory, it represents, within the subjectivity that entertains it, the nature of knowing to that subjectivity—that is to say, it *mis*represents knowing by claiming it to be *inevitably* representational. The bars vanish when, calling upon the reflexive feature of rational awareness, we see that subjectivity's proper cognitive mission is to attain that which is truly other than and independent of that same subjectivity.

The alternative names I have suggested for direct knowing—'rational awareness', 'rational-experiential engagement' (with reality/being), 'rational consciousness/subjectivity'—all call attention to the indissoluble union of two factors that conventional theory of knowledge tends to separate. These names remind us that direct knowing is qualified by a factor of universality and unity on the one hand and a factor of particularity and multiplicity on the other. But the alternative names also insist that direct knowing is not the outcome of applying concepts or other universal items to a preexisting subrational experience, nor the outcome of a formative activity by virtue of which experience as enjoyed or undergone bears the informing/constitutive imprint of rationality. The basic illusion of empiricists, together with the ingenious Kantian remedy for the failure of empiricism, distracts us in our attempt to attain reflexively the functions of mind as they *are*.

Now, as we invoke the reflexive feature of direct knowing to determine what is in fact going on when mind *knows directly*, we notice that the two features of the function we are deploying, rationality and awareness, coexist in mutual support: when the function is deployed reason *experiences;* alternatively, our experience of what we thus attend to is pervaded by *rationality.*

The point may readily be generalized: all supposed mediators of knowing, all items supposed to be starting points for an inferential or constructive process that is alleged to produce, at best, indirect knowledge, turn out to be items that can be directly *known.* Here is an *omnium gatherum* of such items from different but related traditions: impressions, sensations, percepts, sense data (or sensa), ideas, mental images, representations, concepts, universals, terms, predicates, propositions. All such things, whether regarded as things found in mind (consciousness, subjectivity) or as things constructed and sustained by mind, can be at the focus of direct knowing. The function of direct knowing insists on appearing in all its integrity as undergirding any epistemological operation that seeks either to analyze knowing or to build up knowing out of ingredients supposed to be exclusively either experiential or rational.

No activity of the mind, no matter how formal, no matter how designed to exclude any reliance whatever on any experiential factor, is without some reliance on all the bodily particularity of some here-and-now. Conversely, any effort to so isolate the here-and-now as to come upon an experience from which all participation of rationality has been excluded turns out to be merely a misguided exercise in direct knowing. As a consequence, mind ends in contemplating, with the full concreteness of rational-experiential engagement, *items abstracted or formed by mind that then hide the reality out of which they have been abstracted or formed.* Hume's supposed atomic impressions, entirely loose and disconnected, are such deceiving *entia rationis;* so also were the attempts of some of the logical positivists in the thirties of this century to characterize experience without characterizing it by way of propositions—M. Schlick's "Here, now, blue" was such an attempt.

When, by virtue of rational awareness, we attend to temporospatial beings (natural or artificial) that fall within our size range, we achieve what I call *primary rational awareness,* or *primary direct knowing* of those things. We can also attend with the same directness and immediacy to artifacts of another kind, artifacts that are not temporospatial, although most of them are associated with temporospatial entities that are symbols for them. Most important of these artifacts are words, propositions, and languages in general, together with all the momentous structures we form with the help of language; these include not only hypotheses, theories, and mathematical objects but also narratives, poems, and the other imaginative structures produced by the nonliterary arts. I call direct knowing of such things *secondary rational awareness* or *secondary direct knowing.*

This brings us to the second basic function of mind, the one upon which all artifacts depend, the *formative function.* In cooperation with the function of direct knowing, it is responsible for art, technology, and artifacts in general, but just here I pick out only such linguistic artifacts as propositions, together with the bodies of theory we construct with the help of language. Having made such *entia rationis* by virtue of the formative function, we engage them by way of secondary rational awareness, which is just as direct, immediate, and vivid as primary rational awareness. It is secondary only in the sense that whenever we focus upon such things we do so against a background that is always available by virtue of primary rational awareness.

Even when our attention wanders from the background, as you and I might fail to attend to the contents of a room in which we were discussing some problem that concerned us deeply, we can always turn our attention to the background again if there is occasion. Indeed, if what we had been discussing is something very theoretical (say, recursion theory in logic and mathematics or the significance of the red-shift in modern cosmological theory), and if we then turn our attention to the room again (say, to answer a knock on the door), the theoretic "objects" will vanish from our view more completely than the door had vanished while we were talking. The door is there whether or not we attend to it, and to be rationally aware of it we need merely attend to it;

whereas the restoration of recursion theory or red-shift theory *in order that we may attend to it* requires a highly skilled intellectual effort.

Direct though it is, secondary rational awareness nevertheless finds its most important cognitive use in indirect knowing. Direct knowing of theoretical objects that have been formed by mind—such objects as mathematical models of electrons or black holes—is essential for our knowing indirectly the immense variety of things that cannot be known directly, usually because they do not fall within the temporal and spatial size range of our sensory modalities. But indirect knowing of that kind is achieved only by virtue of the leverage our direct rational awareness has on the many concrete temporospatial things that are well within the range of our direct attention. Indirect knowing is one of the glories of human nature, if only because science depends on it: it is the outcome of a complex interplay of secondary rational awareness of theory with the primary rational awareness of commonsense things in which theory begins and in which it is later tentatively confirmed or decisively disconfirmed.

The formative function carries with it a familiar hazard: when we fail to see its distinctness from the function of direct knowing/rational awareness, we may be seduced into thinking that all knowing is indirect, that mind itself can no more be confronted directly than can a quark or a superstring. The originative glance of the mind's reflexiveness, which properly includes both an intense receptivity to what we, as knowers, have in no sense created and an intense receptivity to what is the case about ourselves as knowers, is hindered by the profusion of *entia rationis* generated by the formative function. When such a being of reason happens to be a doctrine or theory about knowing which radically distorts the function of mind, the hindrance becomes disastrous. So it was with what I have been calling the negative philosophical judgment about the powers of the mind. This part of the second movement of the reflexive turn is designed to avert this common misunderstanding by bringing into play the originative but receptive glance of direct knowing.

Although the rational and experiential factors in direct knowing coexist in absolute unity, we do indeed dart back and forth, both in science and in commonsense pursuits, between two quite distinguishable and separable things that are instances of the conventional epistemological distinction between the rational and the empirical: theory/language on the one hand and empirical testing on the other. The distinction between primary and secondary rational awareness allows us to account for this, for in thus moving back and forth we deploy the seamless unity of rational awareness in coping with each of those "poles" of knowledge: we know proposition or theory directly, and we know directly (and do not *merely* experience) that which is empirically relevant to proposition or theory.

⌁

While still attending to whatever is before you, consider now the functions of remembrance and anticipation. To bring the function of direct knowing into play towards my old lilacs or the pile of oak stacked next to my fireplace, I must retain in mind the whole of whatever time I need to take them in: I must hold what is in some sense gone and anticipate what is in some sense not yet here in order to possess myself, in the mode of direct knowing, of what is other than myself. If the act of knowing is seamless in the sense that it is not an alternation between bare conceptual/linguistic activity and bare empirical receptivity, the temporal span in which the act is accomplished is also seamless. This is true of the knowing of a temporospatial entity like a particular lilac blossom seen from the window last May but also true of the knowing of the proposition 'There are lilacs outside my window'. The time taken, short or long, to attend to a blossom or a bush is a unity held together by memory and anticipation; and as you take in and understand a sentence you must hold the first word in mind as you move forward in the sentence in anticipation of its end.

It is interesting to notice in passing that in the Indo-European languages the roots of the words 'mind' and 'memory' are the same or at least intimately related. It is no wonder that as language users we are persistently tempted by the Platonic notion that knowing itself is a kind of remembering. But anticipation is no less fundamental to knowing; indeed, it is fundamental to remembering itself. The temporality we command—that we in fact enact—is future-directed in the sense of being telic: in intending to know something we "reach" for the future as we attend to the thing; and knowing, bringing remembrance with it, is one fruit of that reach. To think of an instant in an act of knowing as caused (under nontemporal laws) by its past instants is to miss what I called in an earlier book the tensive feature of time and so to miss also the temporally tensive aspect of knowing (Pols 1982, 121–22).

In this expository movement of the reflexive turn, we exemplify the much-disputed unity of mind: as we call direct knowing/rational awareness into reflexive play, the functions of intention, attention, remembrance, and anticipation are all ingredient in what we are doing. We attend to direct knowing reflexively, and the telic drive present in all cognitive transactions manifests our intention. Thus, I intend to draw your attention to something I think I see, and you, though perhaps skeptical about what I am saying, intend to see what I am driving at; and even if you find what I am saying to be no help at all, you obviously intend to better your understanding of the relation between mind and body. Conceiving and perceiving also pervade the actualizing of rational awareness, although I have intended to draw your attention to the integrity of their union rather than to their supposed interaction.

Today, when academic philosophy is dominated by a linguistic consensus, the objection that all cognitive activity is intralinguistic and that it is philosophy's peculiar mission to acknowledge that fact must be addressed. The best way to address it is to call attention to a fact about words that seems to have been forgotten: that their primary and probably earliest function is to call attention to concrete things. More generally, one important function of language is to call attention to that which is nonlinguistic or translinguistic. Once that possibility is understood, it is immensely liberating: acts of attention and recognition can then be stabilized in the sense that they can be repeated under the stimulus of the right word.

Liberation came in a rush to the blind, deaf, and unlettered child that was Helen Keller. It came at a time when she could not yet express any propositions at all, let alone the proposition "Things can be named by words." She herself told, many years later, how her teacher held her hand under a pump and let the water flow over her hand, all the while repeating on her arm the sign for water in the sign language devised to teach people in her condition. Suddenly, and without the words to say so, it became clear to the young girl that everything has its name.

The naming function of language is an everyday one, but all the more complex linguistic functions depend upon it, although many and perhaps most contemporary philosophers think it does not exist—does not exist in the sense that the entities named are not known extralinguistically but are in fact linguistic constructs. But this everyday function of language is truly there, and I have been drawing on it in the course of this chapter. The expressions 'rational awareness', 'direct knowing', and 'rational-experiential engagement', for instance, are groups of words, and I have used them to draw your reflexive attention to the function of direct knowing, but only because I could also draw your attention, with the help of other words, to the concrete things in which the power of direct knowing terminates.[6]

The Reflexive Turn, Third Movement: Justification

The third expository movement of the reflexive turn is devoted to justifying and extending the claims made for direct knowing/rational awareness, for this function is central to our reflexive discrimination of the other functions on the list. In the nature of the case, we cannot find a fulcrum outside direct knowing to establish those claims. I turn instead to something intrinsic to the function, namely, the satisfaction that takes place within the knower when rational awareness completes or actualizes itself in something whose being is independent of the knower. As a satisfaction, it must take place in the knower, but, aside from that, it is wholly taken up with the thing known (person, tree, bird, word, proposition, as the case may be), so much so that the only way to bring out its peculiar character is to call it a satisfaction in knowing the other—a satisfaction in acknowledging the known as what it is.

Taking place *in* the knower, the satisfaction qualifies what may be called with equal appropriateness the rational awareness, rational subjectivity, or rational consciousness of that particular knower. The factor of rationality, however, means that the satisfaction is impersonal and universal both as to its "interior" quality and as to its neutral receptivity to that which is other than awareness/subjectivity/consciousness. The complex history of the term 'subjectivity' requires us to restrict the sense we give that term here, and this restriction extends to its correlate, 'objectivity'. We must set aside, for instance, the Kantian notion that the formal structure of objectivity is subjective in origin. With that reservation, the correlate of the satisfaction may be called with equal appropriateness objectivity, reality, actuality, or being, for all of those terms are here understood to convey independence of the function that is satisfied in them. We must also set aside the notion that subjectivity is tantamount to self-enclosure—to isolation from everything that is not a product of the mind.

If the function of direct knowing/rational awareness is real, it is a self-justifying function. This alarming claim must of course be qualified. Direct knowing is certainly not exempt from error in any particular instance, but the reflexive development of the universal factor picked out by the word 'rational' in 'rational awareness' shows the function to be free of the tyranny of the particular instance. Acknowledging yourself to be subject to error, you may nevertheless recognize that the universal factor in the function transcends each instance it is integral with, and so possesses a general authority that is not touched by its failure in a particular instance. Knowing yourself mistaken about just *what* is before you in some particular instance, you nevertheless know that the misidentified thing is other than yourself and so independent of your cognitive act. Confident that your failure can only be defined within the framework of a general competence, you find that you are in a position to try again.

Another qualification is that the authority of direct knowing/rational awareness is by no means at odds with the justificatory procedures that are common in philosophy and science, not to speak of other fields. Although my expository account of the reflexive turn amounts to a discursive justification of the authority of direct knowing, each of the steps has been taken by virtue of an act of direct knowing on my part. As you consider what I have written, you in your turn must address many things that call for successive acts of primary and secondary rational awareness on your part.

The rest of this movement consists of a discursive exposition and justification of the satisfaction that is intrinsic to the function of direct knowing. Both my writing of it and your response to it will depend upon many particular instances of the function held together by the universality that qualifies each instance.

SATISFACTION: PRIMARY AND SECONDARY RATIONAL AWARENESS

The relation between the knower and the known consists in the attainment and enjoyment by the knowing subject of the particularity of the known object—that is, a satisfaction on the part of the knower in just *this* known. Consider yourself as that knowing subject: the otherness of the known object is overcome by you as knower, even while the discrete self-identity of the two beings thus brought together is preserved and acknowledged. Indeed, one component of your satisfaction as knower is your acknowledging, in the very act of taking cognitive possession of the thing known, the utter independence of that object from the function that attains it. But your satisfaction as knower also includes your celebration of the integrity of your own achievement: as knower you have reached out and brought into yourself an awareness of something other than yourself with which you nevertheless acknowledge an underlying affinity. In short, the complex satisfaction is a satisfaction in both the independent particular being of the object and the successful deployment of the function that attains the object.

In the case of primary rational awareness (direct knowing of temporospatial items) the discreteness of the two entities involved, knower and known, includes the distinctness of two temporospatial locations, and this is overcome in the sense that something *over there* is cognitively possessed *right here* in the subjectivity of the knower. Temporospatial location obviously plays a diminished role in secondary rational awareness (direct knowing of theories, doctrines, propositions, imaginative structures of every kind). The mind's role in forming such beings of reason is (or ought to be) part of the satisfaction in its rational awareness of them. Though discrete, though having the integrity of being just *this* proposition, just *this* theory, each of them is understood (or ought to be understood) to be *formed to be distinct* from the knower. Each of them has an inner integrity that must be respected, and as your mind moves through the parts of a theoretical structure to determine their consistency, coherence, and relevance to the matter at hand, their place in the structure has an otherness from yourself as knower that demands rational respect in a way analogous to the demand made by temporospatial objects.

The direct knowing of such *entia rationis* is additionally complicated by the fact that most of them have also a physical instantiation, or embodiment: *Hamlet* exists not only in the mind but on the printed page, the stage, the cinema or television screen; the calculus in the notation of Newton or of Leibniz; the towering achievement of Western music in millions of scores and untold numbers of rehearsals and performances. The satisfaction of rational awareness in such things includes these nuances, though we are seldom explicitly attentive to them. Much of the rest of the story of secondary rational awareness of theory, doctrine, and works of art is interwoven with the complex story of indirect knowing. It is best not to pursue it here, although I have done so in some detail in an earlier book (Pols 1992, 9, 74–77, 146–51) .

In both primary and secondary rational awareness, the satisfaction is unique and *sui generis:* as knower, you cognitively attain the object and enjoy its otherness, but in doing so you allow the object to take possession of your subjectivity/consciousness—so much so that, being filled with the reality of the thing known, you may for a while be utterly forgetful of yourself. In that kind of primary rational awareness in which two persons meet, know, and acknowledge each other, the mutual satisfaction resonates with reflexive intensity.

SATISFACTION: U-FACTOR AND P-FACTOR IN THE KNOWN

One feature of the thing known that is accessible to rational awareness is the tension within the thing known between the factor of unity/universality and the factor of particularity; let us now call them the U-factor and the P-factor respectively. That is not surprising, for the function of rational awareness includes that same tension. On the level of common sense the tension between the U-factor and the P-factor in the known object is easily misunderstood. The *rational* pole of rational awareness reaches out at first to a concept or universal ('tree', 'person', 'triangle', for example) whose ontological status remains a continuing puzzle, and it uses this universal to classify or categorize the particular thing. Sometimes the knower imputes to the universal a causal role in the particular: Plato's Forms and Aristotle's essences are given this role, although in different ways. That tradition associates the universal with the nature of things and the nature of things, in turn, with a deity principle. Our modern epistemological tradition, troubled by a lack of confidence in mind's powers, tends to think of universals as originating in mind and used by mind to project an *apparent* entity—in effect producing a unity and stability where there is in fact only a multiplicity of particular stimuli.

A more persistent intensification of mind's reflexive engagement with the thing known reveals that the unity of the known is intrinsic rather than imposed, and that absolute universality is intrinsic to its unity. Although the rational pole of rational awareness readily attributes universals (predicates, categories) to the particular, it now finds that this can be done only because the formative function of mind is also engaged in the transaction, shattering the

U-factor into a multiplicity of "particular" universals: particular Platonic Forms, particular Aristotelian essences, particular predicates, particular categories, particular species, particular genera, and so on. Such particular universals are beings of reason (*entia rationis*), and when they do not do violence to whatever reality we are dealing with, they are well-founded beings of reason— *entia rationis bene fundata*.

The experiential pole of rational awareness persists throughout this reflexive advance in rational awareness. It is indeed a multiplicity of particulars we are experientially engaged with, and the simultaneous confrontation with the U-factor only means that when we are rationally aware of a particular we are also aware that it shares in a unity that all particulars share in. As I noticed in the discussion of the scientific doctrine of causality, scientists are familiar with one highly abstract version of the U-factor: the laws of nature, regarded as determining the progress of physical systems from one state to the next.

Characterization of the U-factor is always flawed: break it up, for whatever good reason—and the imperatives of science, theology, morals, or art are very good reasons—and we have lost it. Lost it, however, only if we do not acknowledge that one consequence of finitude is that we tend to make things out of the presence of that which we have not made and which is not exhaustively available to us. We have not made the U-factor, and it is not exhaustively available to us. That does not mean that Being in the sense of transcendence is not accessible to our reason: it is as cognitively accessible as my cat on the mat or your book before your eyes. But it is accessible as *transcending* each of the particulars that participate in it. I use that old Platonic word 'participate' deliberately, but with the reservation that we are not considering the participation of a particular entity in a (particular) Platonic Form but rather the participation of a particular entity in a unity/universality to which the formative function of the mind has responded by producing multiple (particular) Platonic Forms.

SATISFACTION: U-FACTOR AND P-FACTOR IN THE KNOWER

As you consider, as knower, the polarity of U-factor and P-factor within the known object, together with your own relation to the known object or objects, you find, within the very operation of yourself as knower, the same polarity you acknowledged within the known object. The particularity of yourself as knower is of course different from that of the known object: attending to yourself as knower—moreover this particular knower, with just this name, this temporospatial situation, this history—you acknowledge that you are distinct from, other than, the known. But to just the extent that it is *rational* awareness that is at issue, you also recognize that your capacity for cognitively assimilating the known is based on a profound affinity between knower and known by way of

6

Mind at the Apex
of a Hierarchy of Causes

The Rational Agent as a Primary Being

Y O U   W I L L   recall that action had an anomalous place in the discussion of the functions of the mind in the previous chapter. On the one hand, the deployment of certain functions—the formative one, for example—can be construed as an action. On the other, a rational action of any complexity calls into play several of the functions on our list, blending them together in the interest of a task of some magnitude. In this chapter, I ask you to complete the reflexive turn by directing your attention not to the functions of the mind as such but, first, to the rational actions by virtue of which the functions are deployed and, second, to the being out of which the actions arise.

The being that is to become the focus of our attention is an agent: a human being that acts—the being in which the action originates, out of which the action comes. As such, it may properly be said to *cause* the action and so provide an *explanation* for the existence of the action. Recall my earlier example of Mozart composing as he played solitary billiards, interrupting his game now and again to write down what he had composed. It is the *being* Mozart that acts and in acting *causes*, on the one hand, the moderately complex set of motions in which eyes, arms, and hands cooperate to strike the cue ball with the cue and, on the other hand, the prodigiously complicated operations of imagining the music and writing it down.

At the commonsense level, most of us will agree that these acts originate in the human being called Mozart. But when the word 'cause' is used, our com-

monsense assent is qualified by a certain puzzlement. Common sense is familiar with the causal relation between the movement of the cue and the movement of the cue ball, but it sees no precise analogy between that relation and
the relation between Mozart and his act. The relation between Mozart and his
act cannot be an instance of causality if causality is understood exclusively in
terms of temporal sequence. To speak more generally: agents do not cause
their acts in the sense in which the prior movement of the cue causes the subsequent movement of the ball.[1]

It is true that the agent Mozart was in existence before he performed the
complex actions I have described, but he is not related to those actions in the
sense in which the movement of the cue is related to the movement of the ball
that follows it. Mozart is *there,* persisting all through the brief time of the action
we are considering: he shapes the act, guiding it to its goal. Before the effect is
realized, he already has the complex effect in mind in a general way, and what
he has in mind is effective in bringing about that effect. The causal order postulated by the scientific doctrine is thus inverted: the telos is effective throughout the sequence of which it is the completion. Moreover, Mozart becomes
permanently qualified as an agent by the complex acts that issue from him, a
consideration that is even more striking if we remember that one component of
those complex acts is the production of music and the recording of it in notation.

It seems clear that if we are to make sense of the causal dependence of an
action on a being that acts, we must manage to unfold a more ample and
many-sided causality than the one which comes down to us from the received
scientific doctrine of causality. I say "unfold" because the key to the matter is
mind's reflexive attention to itself at work. We already have a precedent, in
mind's capacity to achieve rational awareness/direct knowing of something
other than itself, for the complex causality we are trying to attend to.[2] By knowing the other *itself* (and not merely some representation of it), mind *itself* draws
on the U-factor that is intrinsic to both knower and known, and so operates as
a cause in that cognitive transaction in a way that cannot be reduced to the
physical (sequential) transmission of energy, signals, and other "information."

To be sure, a concrete knower and a concrete known are linked by way of
their P-factors as well: they are two particular bodies linked by particular ener-
getic streams. Being so linked, they are bound together in a physical way that
can indeed be given a temporally sequential causal interpretation. But the full-
ness of the cognitive transaction cannot be expressed adequately as a sequential causal transaction. In short, by knowing directly (being rationally aware of)
some concrete thing other than yourself, you operate as a cause in a many-
sided way and moreover reflexively know yourself to be so operating.

The justificatory component of the reflexive turn we considered at the end of Chapter 5 thus also provides a justification of our causal significance as knowers and our cognitive authority to pronounce upon causality in the many-sided sense we are now beginning to unfold. With this, as we shall see, comes a justification of our cognitive access to the many-sided causality of at least some of the other concrete beings we know directly.

In this final phase of the reflexive turn, then, we shall be bringing the topic of causality back to where it properly belongs: with the beings that are causally efficacious. The being that most concerns us is the human being: the being that exemplifies mind at work. But that being is a highly complex one, comprising within its spatiotemporal scope myriads of other beings—from electrons and other subatomic particles up to cells and organs—that contribute to the being of the human being and do so by virtue of their own causal effectiveness.

Throughout this book our task has been complicated by what I have called the negative philosophical judgment about the powers of the human mind.[3] But we are now in a position to overthrow that judgment and reveal mind itself at work as the prime exemplar of the causality that is our theme. The way to an authentic revitalizing of the causality issue lies through a consideration of the causality both known and exercised by the mind of the human being as it achieves rational awareness of the world around it.

Resources for a Reflexive Consideration of the Complexity of Causality

As we take the next steps toward the full actualization of this, the fourth movement of the reflexive turn, we bring with us everything that has been won in the discussion of the functions of the mind in Chapter 5. Let me rehearse briefly the most important of those resources:

- Rational awareness/direct knowing can in principle attain the reality/actuality/being of something other than and independent of the knowing subject. These other things comprise, in the case of primary rational awareness, all the concrete things of the world that fall within the range of our sensory modalities, and, in the case of secondary rational awareness, such beings of reason (in part products of the formative function) as ideas, concepts, propositions, theories, and doctrines.
- The concrete beings so attained are both particular (P-factor) and qualified by a unity that is dependent on a universal unity (U-factor). The U-factor, as noted earlier, can be understood as a universal source of order of which the laws of nature as formulated in any era of history give us a partial expression. Note, however, that the *locus operandi* of the U-factor is absolutely universal: the U-factor is present not just in the

simplest beings that are comprised in a complex being, mandating their assembly into complexes, but also at each higher level of complexity. It is present therefore at the level I have been calling in this book the level of mind itself.

- The formative function is distinct from the function of rational awareness: it does not in principle interfere with the reality-attaining competence of rational awareness. Indeed, when the formative function produces a being of reason such as a concept, proposition, or theory, that being of reason is then a *formed* reality accessible to (secondary) rational awareness.

- Through cooperation with rational awareness, the formative function has access to the U-factor and so can draw on whatever formal order is intrinsic to that factor. That formal order includes logical and mathematical order, but it is by no means confined to that. It is also the source of the order we try to express in morality, art, and science. We appeal to it whenever we refuse to accept as thoroughly definitive any particular inherited code of conduct but try instead to justify the findings of such a code or to reformulate it if it cannot be justified. Any truly creative artist appeals to it in the same way. When Mozart is composing, he is not confined to what is particular to himself, nor does he work only under the formal strictures of an inherited style and tradition. So also with science. Max Planck, in pursuing over decades the line of thought that led from his doctoral dissertation to the construction of the body of theory that yielded the quantum principle, did not work exclusively on the basis of his own temperament and tradition, nor was he confined within the theoretic framework of the science he had inherited.

Cognitive Achievements as Causal Achievements

You have already taken the first step to augment the resources as soon as it becomes clear to you that the cognitive competence of rational awareness and the creative competence of the formative function have a causal no less than a cognitive significance. *Cognitive achievements and creative-cognitive achievements are also causal achievements.* The causes of what mind does are to be found by attending to mind itself. These achievements are adequately explained only if we are prepared to say that *mind itself* performed them—performed them, to be sure, as embodied here in the infrastructure of the being called Mozart, there in the infrastructure of the being we call Planck.

When this next step is taken, it is taken by virtue of an intensification of the cognitive competence of which you have already become rationally aware. That intensification can be regarded as the first step of the next stage of self-

knowledge; as such it amounts to this: to be capable of knowing an independent reality is to be capable of *being a cause* in a sense that illuminates that
achievement. On the other hand, the rubric of self-knowledge is inadequate to
express two features of the achievement: it is a cognitive attainment of real
things that are other than your particular self—the self that is causally responsible for the achievement; it is also a cognitive attainment of the causality
intrinsic to those other real beings—a causality which, in the case of some of
those beings (other persons), is the same as your own causality. Nor is the causal
significance of these other known beings a mere inference from the causality of
yourself, the knower. Direct knowing is competent to find causal significance in
what it knows when that significance is in fact present.

As you take this next step, you are no longer attending to the functions of the
mind as such, but rather to the rational actions in which those functions are
deployed and to the being, yourself, that is the source of those actions. You are
deploying and intensifying the function of rational awareness to grasp the connection between function and action, between action and the being that acts.
You are of course *acting* to do so, but your status as a being capable of rational
action is no mere inference, no mere postulation on the basis of evidence. It is,
rather, a further outcome of your deployment, as rational agent, of a function
whose reality-attaining competence includes also this reflexive attainment of
your own status as a primary being and the status of others who are also primary beings.

The Hierarchy of Causes Redux: Mind Itself and Its Infrastructure

Your rational life is by no means confined to such convoluted reflexive returns
of cognitive competence upon itself. Most of it is lived in a world in which that
same competence is exercised in finding out the "particular go" of the many
complex and concrete things we find there. One of these things is that extraordinary hierarchy of complexity, the human body. We return, then, to where we
began in the Introduction. But, having now some assurance of the competence of
mind to know items of the infrastructure, some assurance of the causal/explana-
tory significance of ourselves as rational primary beings, and some assurance of
the causal/explanatory significance of items in the infrastructure, you can begin
to see how mind itself stands at the apex of a hierarchy of causes.

Because you have brought the functions of mind itself into play to determine
their own nature and authority, you can now deploy the functions again with
some confidence in that authority. But this time your target is the dependence
of those functions on your own causality as the apex being of a causal hierarchy—the apex being of an infrastructure that defines the embodied state of the
human mind.

When you do so act rationally, you stand at the apex—indeed you *are* the apex—of a pyramid of causes made up of untold myriads of entities/beings, each of which is the apex of a smaller pyramid of causality. As you deploy various functions in an act, say, the act of reading the words on the page before you, you also deploy your causality—your power of determining something— down through the multiplicity of the pyramid. As I said in the Introduction, the metaphor of a pyramid should not be taken literally. There is only one item at the top, but it is not located at the top in the sense in which the top block of a pyramid made of blocks sits on the very top and so can be pointed out there. You are at the top in the sense that you are only *one*, while each level below you is made up of *many* items; but on the other hand your own and very particular self-identity cannot be discussed, let alone established, without reference to the multiplicity of beings below you in the pyramid.

Some of these beings—organs, for instance—are easy of access; some, however, are knowable only by way of theory, and the lower down we go in the pyramid (the infrastructure), the more theory predominates. But, as we saw in Chapter 5, the indirect knowledge we get by way of theory is the outcome of the cooperation of primary rational awareness, the formative function, and secondary rational awareness. It is thus your capacity for rational awareness that makes sound theory about the infrastructure possible.

The multiplicity of the infrastructure baffles the imagination. By today's usual estimate there are some 10^{11} neurons in the central nervous system, and some 10^{13} cells in the body, that is to say, a trillion trillion (American style, in which a trillion is a million million). So it goes downward in the pyramid, the multiplicity growing more preposterous with each step of the hierarchy: by one authoritative but probably conservative estimate there are some 10^{15} macro-molecules in the central nervous system, and the central nervous system is but a relatively small part of your pyramid.

Each being in the multiplicity of your pyramid contributes its power "upwards" by virtue of its own causal authority (its own determining power) to the causality you exercise in acting rationally. Massive as that causal contribution is, you in turn exercise your own causal authority "downwards" over each of these entities, determining some of its career in a way sometimes incidental, as when by moving your hand you move the molecules that are parts of it, and sometimes profoundly fundamental, as when by saying or deciding this rather than that you produce this rather than that pattern of electrical firing in your central nervous system. You do not produce it by producing a physical process that feeds into the pattern of firing in a $C{\rightarrow}E$ sequence of the kind discussed in Chapter 3. You cannot deploy any mental function without the support of the very pattern of firing in which your mental function is causally effective in a nonsequential sense.

What is determinable is open to determination by what is in some sense distinct from it, and you at the apex of your pyramid are in some sense distinguishable from the biochemistry of your brain, although we have only begun to examine that sense. You may decide to study calculus or Chinese or Swahili or all three, and the study or studies undertaken will make a difference in your brain, as indeed will the act of deciding itself. But what is open to determination need not be pervaded by some absolute indeterminacy or chance in order to be thus open.

Many, if not all, of the entities in your pyramid—a single neuron, for instance—deploy their own causal authority down through their own pyramids, although no doubt theirs is a less momentous determining power than your own, since your own power can in principle complete itself in moral decisions and actions and can, in some happy circumstances, issue in a work of art or science. These beings within your own pyramid are related to the items within their pyramids in a way analogous to the dependence of your self-identity on the particularity of your own pyramid. Thus, the cell determines "downwards" at least some of the career of a macromolecule that contributes to the cell's being that particular cell; and, as in your own case, the items in the cell's pyramid contribute causally "upwards" to the determining power exercised by the cell. The point may also be expressed by saying that the unity of nature manifests itself in the self-identity of the cell no less than in the self-identity of the cell's components.

The expressions 'an entity' and 'a being' are truly ubiquitous: we apply them to electrons, to the particles in the atomic nucleus, to atoms, molecules, one-celled organisms, animals, cities, schools, corporations, planets, stars, galaxies, and indeed anything that we can single out by its apparent unity from the rest of the environing world. The beings I have been considering so far make up only a certain class of that larger number: those that possess a causal authority analogous to the kind I have been discussing. I have already suggested that such entities are best called primary beings, or primary entities. Because this "vertical" causality exercised downwards and upwards in a hierarchy is intrinsic to the primary being that exercises it, I find it helpful to call it *ontic causality* or *ontic power*.

Some of the beings in your pyramid and in my own—certain molecules, for instance—are detachable: they had earlier careers and will presumably outlast either of us. Some other beings—our organs, for instance—seem so close to our own self-identities that we find it plausible to think that we are composed of them: they are more or less stable structures just as we ourselves are more or less stable structures of which they are parts. These structures are telic in at least this limited sense, that if they fail to carry out their functions, higher functions also fail. The heart is a pump, and as such it contributes to the environment necessary to sustain the brain, which in turn sustains, though it may not entirely account for, the functions of mind. Certain functions are thus necessary for, although they do not necessitate (are not sufficient for) the higher functions—a distinction that can be found as early as the passages from Plato's *Phaedo* and *Timaeus* I discussed in Chapter 1.

Some organs can be transplanted, but they depend in the end on some organic pyramid, just as that pyramid depends on them. Certain other beings—organelles, for instance—have careers only within our cells, and they are gone when the cells are gone. Some of the primary beings within our pyramids seem to have act-like functions, or in any event we are tempted to use the notion of action as a metaphor when we talk about them: so it is, for instance, with the organelles that bustle about within a cell transporting energy from one place to another. Not all of them can be said to act, not even metaphorically; but all of them can plausibly be said to have functions within the pyramid of entities/beings to which they belong.

Rational action always calls into play the function of the mind called rational awareness/direct knowing, together with the range of other functions discussed in Chapter 5. As I remarked earlier, the Greek word for function is a plain one: *ergon*, the ordinary Greek word for work. Contemporary philosophers who are specialists in the study of mind tend to rely on the notion of a function and to dismiss as unscientific the notion of a being that performs the function. What I am calling a pyramid of beings is often discussed in philosophical journals as a set of functional levels, and the reader gets the impression that discourse in terms of such levels is more respectable than discourse in terms of the beings that are the bearers of the levels—the beings that perform the various kinds of work that lead us to talk of functions.

Still, there is something persuasive about the notion of an entity that is capable of various functions: it would appear, for instance, that I perform the function of writing as I produce this paragraph and you perform the function of reading to take it in. We might go even further: to speak of either of us as having a mind is to think of us as beings capable of exercising a number of functions grouped together under the general rubric of mind.

The Complex Temporality of "Vertical" and "Horizontal" Causes

Consider now the power you exercise in performing any of these functions. In one sense, that power is nothing if not temporal: it takes time to think and to plan; it takes time to perform an act like writing or reading a paragraph. It also takes time for a being lower down in the pyramid to carry out one of its functions: it takes time for a neuron to be excited to the point of firing by other neurons that are in synapse with it; it takes time for potassium and sodium ions to move through the ion gates in a neuron's wall; and it takes time for the neuron to fire, rolling a changing electrical potential along its axon. Some of these temporal happenings have a later effect at other functional levels: if the heart stops, brain damage will happen later; if someone decides to take up cross-country skiing, improvement in the cardiovascular system will occur later.

But it is otherwise with the causality you deploy downward in your pyramid in the course of such a rational action, and it is otherwise with the causality each being in that pyramid deploys upward in support of that action. *That* causality is the exercise of a nontemporal power—nontemporal at least in the sense that in deploying it you do not do something somewhere or other whose impact or influence in the multiplicity of the level below you (say the neuronal level) *only* appears there *after* you deploy it. You do not think and afterwards produce electrical patterns in your central nervous system: you think by virtue of the patterns to which your thinking contributes. So also with the support given your act by the neuronal level: each neuron does not do something whose impact or influence only appears afterwards in your thought.

All this is not to deny the temporality of the functions performed by the level of your mind or the temporality of the functions of the neurons that fire in the course of your action; it is rather the relation between those two developing levels that I call to your attention and assert to be nontemporal. It is, to be sure, a curious relation, for we cannot distinguish in any absolutely unchallengeable way the two things so related. I cannot, for instance, point first to one of the things, you, and then to the other, your body, as I could in the case of two physical objects I asserted to be related. The difficulty, of course, is that in some sense you at the apex of your pyramid are identical with the multiplicity of functioning items in the pyramid that you also are. Your self-identity is asymmetrical: on the one hand, it is a One, on the other, it is a Many; yet the two factors are so intimately united that the word 'relation' does not do justice to their union. It is the self-identity of a being that we are concerned with, and the self-identity is asymmetrical in the sense that the causal status of the One and that of the Many are distinct.

Most of the time we take for granted what I am now trying to call to your attention: that the One and the Many—both of which, taken together, are *you*—are involved in a power relation that is fundamental to your moral nature and to your status as a rational being. This power relation is part of what we mean (or should mean) when we talk of self-integrity and self-control and take them to be essential to our true self-identity. As beings, you and I dwell within the constraints of our multiplicity in the interest of our own ideal unity, a unity whose demands we do not entirely understand, perhaps because (as I argue later) it is not something that belongs exclusively to any of us. We are not, that is, wholly responsible as *particular* beings for the causal authority of the One on which our self-identity depends.

Immanent and Transeunt Causality of Primary Beings

If you, as the apex of a pyramid of beings, exercise a nontemporal power within and by virtue of your neuronal system when you think or otherwise act rationally, if that multiplicity of neurons exercises a nontemporal power without which your rational action could not take place, we can say that you, who are thus both One and Many, have a causality that is *immanent* in you. The same thing is true of any of your neurons: it possesses an immanent causality exercised downwards in its own pyramid and upwards in contributing, as one of your Many, to your rational act. As apex being, you have an immanent power that is exercised both within and by means of the vast multiplicity of beings in your pyramid: if your power constrains each of the beings in your infrastructure, it is also constrained by the power immanent in each of them. Because this mutual constraint of One and Many is the source of our particular self-identities as primary *beings,* the alternative names I suggested for immanent causality—'ontic causality' and 'ontic power'—seem to be happy ones. Causality is, after all, merely the being of a thing regarded in a certain explanatory light.

Recall, however, an earlier caution about functional levels: such levels exist only in the sense that there are primary beings capable of deploying them—deploying them, that is, by virtue of deploying ontic power downwards in their pyramids. If we think of the mind-body relation as consisting in the relation between two functional levels—say, consciousness and the neuronal—each ontologically complete and independent in itself, each capable of acting on the other, we are back with a familiar and untenable dualism. We are caught in the trap of regarding *consciousness* as exercising ontic causality on the neuronal level and in return being supported by that level. The point of ontic causality is more subtle than that: the point is rather that the unity of the being in question *achieves* consciousness of things in its world by virtue of deploying its ontic causality in the neuronal level—and of course receiving support from the neuronal level. Consciousness is the outcome of ontic causality, not an independent source of it.

The ontic causality intrinsic to you is also deployed in the world about you: by virtue of your ontic power you affect other beings in the world about you; you cause happenings in that world. As medieval writers used to say, you deploy *transeuntly* the causality that is *immanent* in you.[4] Transeunt causality is more accessible to our commonsense attention than ontic causality, so much so that many people take it to be the only mode of causality. (They do not, to be sure, often use the qualifier 'transeunt'.) The reason for the accessibility of what I am calling transeunt causality is plain enough: transeunt causality involves a temporal sequence in which two distinct items can be discriminated—one in which the power originates (you or me), the other on which the power is exercised (some item in the world around us). Your rational acts thus necessitate or determine happenings outside the immanent tension of One and Many that is you; so do mine: my act of writing has contributed to a printed page that is now before you. Our confidence in the reality of transeunt causality, however, is soon eroded if we are persuaded that we can find no ontic causality/power out of which it springs. That, in a nutshell, is the story of Hume's attack on what he took to be the notion of causality.

The Ambiguity of the Self-Identity of Primary Beings

How characterize an entity, such as yourself, that is both One and Many? If we try to locate the One we cannot do so except by pointing to the person—pointing, that is, to a physical presence that is also the pyramid understood as a Many. If we insist that we are indeed locating the One, because the being we are discussing is a One *in* a Many, we must at least concede that it is not a physical pointing that does the job but rather a "pointing" in which the mind participates. The One, which we insist is in some sense the apex being in question, is alleged to preside over the Many in which and by virtue of which it achieves its purpose; and in almost all cases this achievement results sooner or later in some transeunt effect on the world about it. In the case of the two of us we may take the purpose of a One-in-Many to be what I am saying to you in the course of writing a sentence or what (if we were conversing) you might say to me in response.

If the attention of the mind lapses, and we point to such a One as a human being by way of a physical gesture, we may take the One to be merely a name for an assemblage which is a Many—this particular Many that we take as a One. The One as a center of causality has vanished. If we then call into play once more the attention of our embodied mind, the One reappears. But its reappearance is problematic: the more we focus reflexively on it and on our cognitive response to it, the more clear it becomes that the One cannot be adequately understood in terms of the notion of mere particularity. The One reappears as a power that is a factor in the particular you or the particular me but does not belong to the particular in the sense in which the particular's Many does.

The One that reappears is intrinsically universal: it is the same One that appears in each being of the bewildering Many that makes up your pyramid or mine, from the apex being—the person I take you to be or the person you take me to be—down to the smallest particle which theory allows us to postulate within the pyramids. In philosophical jargon, the One that is intrinsic to each particular being that possesses ontic causality is also the transcendent One. It is often taken for granted that the notion of transcendence belongs with religion rather than philosophy or science, so it is important to remind you that the transcendent One is acknowledged by science, although under a different name. The belief in the unity of the laws of nature is a belief in a transcendent One, at least if that unity is thought to have an ontological status by virtue of which it determines the structure and career of each physical entity and the structure of time and space as well. But we now reinterpret the ontological significance of the laws of nature in the way suggested in Chapter 4: we now suppose the unity of the laws of nature to be a highly abstract version of what I am here calling the transcendent One.

Alternatively, we suppose that there is a transcendent/universal aspect to the immanent/ontic causality that is intrinsic to persons at one end of the scale of complexity and to such entities as electrons at the other. In response to the claims of microentity reductionism, we have dismissed the claim that the (transcendent) nature of things has its *locus operandi* only in the microentities of the base level. For that we have substituted the claim that its *locus operandi* is in the apex of each primary being from the most evanescent particle to such highly complex beings as Newton and Mozart.

The Key Role of Rational Awareness in the New Hierarchical Causality: The U-Factor Again

The point just made may be illustrated by taking our examination of rational awareness in Chapter 5 as a precedent to make a negative point about the scientific doctrine of causality and a positive point about hierarchical causality.

The negative point is that if rational awareness does what it purports to do— if it reaches out and actualizes itself in something other than itself—then it cannot be adequately accounted for in terms of the scientific doctrine of causality. Suppose the physical system (*PS*) we are considering embraces both knower and known, and that we call a state of the *PS* just before an act of rational attention $PS(s1)$ and the state that coincides with an act of rational attention $PS(s2)$. In that case, the part of $PS(s2)$ which is the knower will include a causal contribution (causal in the sense of the received scientific doctrine) from that part of $PS(s1)$ which is the known, and the state of the knower qua effect will thus be distinct from the state of the known qua cause.

The distinctness might take the form of the known being the *original* and part of the state of the knower being some sort of neural *representation* of the known. This representation (or mapping, or coding) might well be biologically useful: by virtue of it the knower might react to the known in some biologically profitable way. But its usefulness is irrelevant to the question before us, which is whether the distinctness of effect from cause (the scientific doctrine defining the only sense of causality operative) can be identified with the much more intimate relation between knower and known in which the first is rationally aware of the second. To put the matter in a different way: rational awareness of x cannot simply be an effect of x in the sense of 'effect' postulated by the scientific doctrine of causality. It does not matter whether we regard the conscious component of the relevant part of $PS(s2)$ as identical with that part, as epiphenomenal upon it in some noncausal sense, or as a nontemporal effect of it, that is, an effect of an effect.

The positive point is that if rational awareness does what it purports to do, a temporal and sequential causal relation can play only a partial role in its achievements. The temporal/sequential relation I have in mind is the one I called transeunt causality earlier in this chapter and $C{\rightarrow}E$ causality in Chapters 3 and 4. But my main point is that a temporally sequential causal relation, however understood, is just part of the story of the relation between rational awareness and the known entity or situation towards which it is directed. The causal structure of this most basic of our cognitive achievements comprises also both the ontic (nontemporal) causality exercised by the apex being (the knower) and the ontic (nontemporal) causality exercised by all the beings of the knower's infrastructure.

The *sui generis* relation between knower and known called rational awareness originates in the knower (the apex being) in the course of an act of attention on the knower's part. Rational awareness takes place in time, but it takes place by virtue of the nontemporal union of the ontic powers of the apex being on the one hand and the beings of the infrastructure on the other. But transeunt causality also has its place in the achievement of rational awareness. All sorts of transeunt causal processes link the knower and the known, but the one that concerns us most is the complex transeunt transaction in which that part of the infrastructure of the known we call the central nervous system is affected by the thing or situation known.

This transeunt influence qualifies the infrastructure (P-factor) of the knower, and hence also qualifies the ontic contribution made to the apex being by the ontic power of its infrastructure components. But the outcome of this complex causal structure is not a cognitive response on the part of the apex being to its own infrastructure as that has been transeuntly affected by the known. It is, rather, a cognitive response to the thing or situation known: *the knower achieves rational awareness of the known.* And that independently existing known is other than the knower and also other than the transeunt causal effect the known has produced within the knower's central nervous system.

But, you may ask, how can this happen? If when you raise that question you are demanding some account of a *transeunt* transaction between the infrastructure elements that make up the central nervous system on the one hand and the "mind itself" of the knower (the apex being) on the other hand, you are missing the point of this present account of the relation between knower and known. Transeunt causal transactions ($C{\rightarrow}E$ transactions) do not take place between an apex entity and the entities in its infrastructure. And indeed, if they could take place, the outcome would have to be a transeunt effect of the transeunt effect that the known entity or situation had produced in the central nervous system of the knower. No: if rational awareness of the independent entity/situation I have been calling the known is in fact achieved, then one factor in the nontemporal "transaction" that is vital to bringing it about is to be found in the relation between the ontic power of the apex being and the ontic power of its infrastructure elements.

That, however, is only part of the story: the apex being (the knower) cannot wring the independent known out of what is *not* the independent known; and the knower's infrastructure is *not* that independent known but is merely in transeunt causal connection with it. So besides this important nontemporal ontic "transaction" between the apex being and its infrastructure there must be something else involved in this most basic of our cognitive achievements.

If rational awareness of the independent known is indeed achieved, what is decisive in the achievement is the U-factor intrinsic to the apex being that is the knower together with the U-factor intrinsic to the independent being that is the known. I say "together with" because these two U-factors are not unqualifiedly two, but rather two instances of participation in the same U-factor by two distinct entities, each individualized by the vast multiplicities of their two infrastructures (P-factors). Rational awareness, if it exists, rests on an affinity between discrete entities that goes back to an ordering power they share in. How all this "works" we do not know and probably cannot know, because the contribution of the (transcendent) U-factor is nontemporal, and it is as intimately concerned in the known as in the knower.

We are, then, driven back once more on mind itself, for only mind itself, in each of us, can determine what the functions of our mind can and cannot do. And since even those of us who are materialists constantly make determinations that this or that *is truly the case,* it seems that all of us acknowledge, at some level of discourse, a reality-attaining competence that belongs to mind itself.

How Reasons Can Be Causally Effective

The transcendent One in some particular Many really *is* the being in question—you or me. The unity of a particular primary being is elusive only when we suppose ourselves to be hunting for a One that is the exclusive property of the particular being whose Many we are attending to. The self-identical actuality of each being that possesses immanent, or ontic, causality thus consists in an asymmetrical union of a universal One with the particularity of a Many on which it confers unity. Any being that is the apex being of a pyramid of many beings is a *particular* being by virtue of that pyramid, but it is *one being* by virtue of a One that is not unqualifiedly particular to it. It is the Many intrinsic to any such being which makes it particular; it is the One intrinsic to it which makes it not merely a particular. We are such beings in all our occasions but most vividly so in the exercise of our mental functions.

The temporality of each being—the apex being of our pyramids, the apex beings of each identifiable pyramid at a lower level—thus pivots on a nontemporal factor that is both one and universal. Each apex being finds itself in a temporal environment—that of the Many of which it is the One—and each apex being also makes its own contribution to the complexity of time as it carries out its functions in and by means of its Many. The phrase 'its own contribution', however, must be qualified, for we are not considering a radical plurality of Ones but a plurality of Ones that share in a universal (thus transcendent) One. Because of that sharing, a primary rational being can make reasons causally significant factors in its actions—reasons in a sense broad enough to comprise the distinct modes of order that define and pervade such fields as logic, mathematics, science, morals, and the arts. Reasons (so understood) are causally significant in the sense that it is only by attending to the *reasonable* structure of the actions and their by-products that we can understand how they came into being. Reasons emerge gradually from the mind of the agent in a linked series whose significance concerns the whole of the action, the whole of any work that results from the action. The causal significance of reasons is thus obviously telic. That significance is also in part impersonal, for to emerge from the mind is to emerge from a matrix that exists by virtue of the union of the U-factor and the P-factor. To be sure, the coming into being of actions is also dependent on the infrastructure of the rational agent's acts, but the union of the apex being and its infrastructure is so close that the unfolding of reasons in the ontic level of action is also the making of reasons effective in the infrastructure.

In one sense, these reasons are airy nothings, but in assenting step by step to them as the order of its action develops, the rational primary being makes these airy nothings effective not just in the logic of the (transeuntly effective) action but also in the physical systems that subserve the actions as infrastructure components. The rational being has, so to speak, one foot in the normative and one foot in the physical: it is an *embodied* rational primary being, and everything it does in the sphere of reason and norms it does also in the physical. Its goals, together with the reasonable steps to achieve them, are thus "real causes" in the sense intended by Socrates in the passage from the *Phaedo* quoted in Chapter 1, and so they "operate" in the infrastructure in a nontemporal way that has some resemblance to the way laws of nature are sometimes said to govern any physical process that takes place in a physical system.

The Self-Identity of Primary Beings and the U-Factor: Participant Powers

This leaves us with a difficulty about the explanatory value of each particular primary being. On the one hand, causal explanation terminates in each of the particular beings that exercise immanent causality and so are capable of exercising transeunt causality in the world about them. On the other hand, no such particular is an ontological absolute. There is a dependence intrinsic to any particular primary being's status as a One: it is a particular One, a particular ontic power, by virtue of the universality of the One in which it participates.

In an equally important sense, then, causal explanation does not terminate in each *particular* One that exercises immanent causality (ontic power) but rather in the One in which it participates. Primary beings are ontic powers, but only by virtue of being participant powers. This is a mystery in the sense that we cannot say why it is so, but must simply acknowledge our dependence and acknowledge also that dependence does not exonerate us from the obligation to act as though everything depended on us as particulars. In an odd sense everything *does* depend on us qua particulars, for none of us is a mere particular: there are no mere particulars.

You and I, for the moment reader and expositor engaged in related discursive activities, can resort once more to philosophical jargon and say that the U-factor, which is intrinsic to each particular, has a causal no less than an epistemic significance. Or we can use another formula to draw attention to the source of our cognitive competence: the integrity of the union of the U-factor and the P-factor in ourselves is the *reason why* we succeed in the commonplace marvel which is the cognitive attainment of some primary being other than our particular selves. In short, the crucial feature of direct knowing is also a causal feature: direct knowing, which is a function of a causally significant One embodied in a causally significant Many, is in that sense self-explanatory.

There are other ways of drawing on the attention-calling power of language to facilitate the full actualization of the reflexive turn—other ways of unpacking, in an expository and discursive mode, the extraordinary simplicity-incomplexity of direct knowing. Here is one: when as knowers we attain the reality/being of some primary being other than ourselves, we attain at the same time the causal authority of that other in *its* hierarchy and (in the reflexive turn) the causal authority of ourselves as knowers in our particular hierarchies. Here is another: the epistemic feat of achieving primary rational awareness of certain other beings includes their status as immanent causes (ontic powers); but so also with our cognitive attainment of the being of our particular selves: we know ourselves to be immanent causes. Moreover, in virtue of the U-factor, there is never an absolute separation of the case of the knower as immanent cause from the case of the known as immanent cause.

Overcoming the Negative Philosophical Judgment about the Power of Mind Itself

With this phase of our reflective turn we overcome the negative philosophical judgment about the power of our minds which has gradually developed in the course of the epistemological era that is coincidental with modernity—including that last jaded period of modernity which now masquerades as postmodernity. The negative judgment led to the dismal conviction that we can know neither other beings nor ourselves directly and so cannot know any causal significance they may have. It contributed to the gradual discrediting of a complex and nuanced doctrine of causality which, whatever its limitations, left some explanatory room for morality, art, and religion. It left us with a scientific doctrine of causality for which it is impossible to provide a satisfactory philosophical justification. To speak more precisely, it condemned us to oscillate between the conviction that the scientific doctrine is a thoroughly adequate replacement for all the older views of causality and the suspicion that it is a useful pragmatic device whose writ must be carefully circumscribed. It is time to set aside the negative philosophical judgment once and for all.

Notes

Chapter 1. Plato and Aristotle on Mind, Soul, and Causality

1. Plato uses the terms *on* in *Sophist* and *ousia* in *Timaeus* in discussing what I am here calling the Form 'being'. All the language here in *Timaeus* makes it clear he is talking about the Forms—hence about Being and the Form 'being'. F. M. Cornford—great commentator though he is—produces in my judgment a false emphasis by choosing the translation 'existence' rather than 'being' for those terms. His commentaries appear in the references under Plato's name.

2. Another related passage, *Metaphysics* 1038b19, makes it clear that the universal form—the form entertained as something intelligible and definable—is not *ousia* in the full (primary) sense proper to the actuality of informing form.

3. The term 'essence' (*essentia*, from *esse*, 'to be') has an illustrious history in Western philosophy in translations of Aristotle and in many kinds of philosophy that owe something or other to the Aristotelian tradition. But it is also used in contexts that are clearly Platonic rather than Aristotelian. Thus Descartes, in discussing what he can infer from his idea of a triangle, uses *essentia* interchangeably with *forma* and *natura* to speak of what seems to be a Platonic Form of triangle (*Meditations*, V, para. 5). See Chapter 2.

4. Some of the tangles in the central books of the *Metaphysics* can be resolved by turning to *On the Soul* and *Nicomachean Ethics*, because soul as (actual) essence of a primary *ousia* is the main topic of those two books.

Chapter 2. Descartes's Dualism and Its Disastrous Consequences

1. The expression 'principle of sufficient reason' seems to have been invented by Leibniz. The first appearance of that expression is in article 32 of his *Monadology* (Leibniz 1714/1898), but at the end of that article Leibniz refers us to article 44 of his earlier *Theodicy*, in which he had used the expression 'determinant reason' instead of 'sufficient reason' in a formulation of the same principle; in the same context 'cause' is used as an alternative to 'reason' at one point (Leibniz 1710/1952/1966). Robert Latta, in his book-length introduction to his pioneering translation of some of Leibniz's work (Leibniz 1714/1898, 160–63), accurately describes Leibniz's principle of sufficient reason as a generalization and clarification of the causal principle of Descartes I quote in the text. To be sure, Descartes's causal principle is not the only precursor of Leibniz's principle of sufficient reason. The notion that everything has a cause and that the cause must be adequate/sufficient to produce it is a commonplace of the Aristotelian tradition. The earliest clear formulations of such a principle are probably those of Plato in *Philebus* (26E) and *Timaeus* (28A).

Chapter 3. The Received Scientific Doctrine of Causality

1. In earlier books, I have sometimes discussed this version of causality under the heading of $C{\rightarrow}E$ causality, but my views on the status of the laws of nature are different from those of writers

who accept the received doctrine as an adequate version of causality (Pols 1975, 1982, 1992). I return to this topic again in Chapter 6.

2. There are other cases in which the gradual assembly of necessary conditions does not appear to amount to sufficiency. We have looked at a classic case in Chapter 1: Plato's examination of the reasons for Socrates' remaining in prison to accept the sentence of the Athenians rather than taking advantage of the evident wish of the authorities to allow him to escape (*Phaedo* 98–99). But the whole point of that famous passage, which I have examined in greater detail in an earlier book (Pols 1975), is that the capacity for understanding reasons and the capacity for moral choice cannot be adequately understood in terms of a causal scheme intended for the analysis of a purely physical process. Just here, however, we are concerned with a received scientific doctrine of causality that grew out of a determined effort to analyze physical processes without regard to such matters as moral purpose.

3. Because I want to make the notion of law central to my contrast of the scientific doctrine of causality with the kind of causality I take to prevail in causal hierarchies, I run the risk of seeming to ignore the extreme complexity of the scientific enterprise. I am well aware, however, that laws make up only part of the body of theory a working scientist deals with. In speaking of a body of theory I am using the word 'theory' in one of its more comprehensive senses. A *body of scientific theory* includes a great many things, and the following catalog of them is far from complete: definitions, assumptions, transformation laws, theories (in a narrower sense than in 'body of theory'), hypotheses, models (both physical and mathematical), a range of mathematical objects and techniques, and finally laws of nature—or rather formulations of them. That last qualification is important, for the notion of the laws of nature is not exhausted by the collection or system of currently accepted laws. As Laplace clearly recognized, that notion is also an ideal for science, and it mandates a distinction between today's formulation of a law and the real law scientists are trying to express in that formulation.

Chapter 4. Mind and the Scientific Doctrine of Causality

1. See the discussion of beings/entities and of the corresponding Greek terms in the Introduction and Chapters 1 and 2 of the present book. The same expressions are brought into service again in Chapter 6.

2. See the Introduction. The notion of an apex being is developed in more detail in Chapter 6.

Chapter 5. Mind on Its Own Functions

1. Most academic philosophers would speak here not of beings and causality but rather of substances and their causality. I have given some reasons, in the Introduction and Part One, for finding the notion of substance inadequate (at least in modern times) to the Aristotelian notion of *ousia* that lies behind it. I develop those reasons further in this chapter.

2. The term 'conscious' in somewhat its present sense was current in English for the greater part of the seventeenth century: it occurs in a literary setting in Massinger (1632) and Milton (1669), for instance (*OED*, 2d ed.). Perhaps the first occurrence of the term 'consciousness' in a philosophical setting is in a book by a Neoplatonist, Ralph Cudworth's *True Intellectual System of the Universe* (1678); but Cudworth was acquainted with Locke, and Locke was engaged from as early as 1671 in the discussions and writings that led eventually to the publication of his *Essay Concerning Human Understanding* in 1690, in which he makes considerable use of 'consciousness' (II, xxvii, 9–10). It therefore seems probable that the term was current in philosophical circles in the 1670s.

3. See also the detailed discussion of all these matters in Pols 1992.

4. Though Hobbes's materialism has led him, according to Descartes, to choose the "concrete" word 'subject' in speaking of the mind, Hobbes is right, Descartes concedes, in saying that we cannot conceive of an act without its subject (Descartes 1991, 2:123–24).

5. Kant's discussion of transcendental subjectivity occurs in the latter part of the *Critique of Pure Reason*. There he is concerned to make it clear that transcendental subjectivity is merely the subjective condition of objective (synthetic a priori) knowledge, and that to invoke transcendental subjectivity is not to pretend to know oneself as a thing in itself. In the sense in which Kant intends the term, all particular knowing subjects share in transcendental subjectivity, and so all of them must understand the world in the same universal and necessary way. Another way of making the same point is this: transcendental subjectivity provides the categories of the understanding and the forms of (sensuous) intuition that confer upon the raw material of sensation the formal constitution we call objectivity.

6. Although the topic of reference has been endlessly debated within the linguistic consensus, there is no problem whatever about reference if we confine ourselves to the drawing-to-attention feature of language in the case of direct knowing. There is a profound problem, however, when analytic philosophers dismiss everyday, direct-knowing reference as an illusion and focus instead on reference to theoretical entities and to the linguistic items that do indeed enter into our reference to such things. In the supposed absence of direct-knowing reference, these philosophers call on the discipline of semantics to do whatever can be done to establish the relevance of words to things from within the supposed linguistic enclosure. It is not surprising that they find the reference problem intractable in that setting. But indirect knowing is parasitic on direct knowing, and not all of the many problems that surround indirect knowing are relevant to direct knowing.

Chapter 6. Mind at the Apex of a Hierarchy of Causes

1. In chapters 3 and 4, under the rubric of the received scientific doctrine of causality, I consider rigorous versions of the doctrine that the cause occurs before its effect. But there were various commonsense versions of that mode of causality long before the modern scientific version was developed. The scientific version is an attempt to develop rigorously what is already present in common sense. It was, however, one conclusion of Chapters 3 and 4 that the development of the scientific doctrine, coupled with the prevailing epistemology, erodes our confidence in common sense.

2. By "something other" I mean both concrete things and things such as ideas, propositions, and theories that mind has formed on the basis of its intercourse with concrete things.

3. To sum up that judgment again: it represents the mind as incapable of knowing directly (being rationally aware of) beings other than itself and incapable of knowing the causal relation directly.

4. Thomas Aquinas, for instance, contrasts immanent action (*actio manens*) with transeunt action (*actio transiens*) and immanent operation (*operatio manens*) with transeunt operation (*operatio transiens*) (*Sum. Theol.* 1a, Q. 18, 3, ad 1; 2 *Contra Gentes* 1 [Deferrari et al. 1948, 14, 770]). As it happens, Aquinas is talking about God in these passages, but the distinction 'immanent/transeunt' makes good sense in the human context as well.

References

Aristotle. Works of Aristotle are cited in the text by the standard (Bekker) pagination, which usually appears in the margins of good modern editions. The following editions of certain of Aristotle's works were used by the author but are not cited in the text.

——. *Categories*. Trans. H. P. Cooke. In *Aristotle: The Organon*, vol. 1. Trans. Cooke and H. Tredennick. Loeb Classical Library, 1938.

——. *Metaphysics*. Trans. H. Tredennick. In *Aristotle: Metaphysics; Oeconomica & Magna Moralia*. 2 vols. Loeb Classical Library, 1933, 1935.

——. *Metaphysics*. Trans. R. Hope. New York: Columbia University Press, 1952. Paperback reprint, Ann Arbor: University of Michigan Press, 1960.

——. *Nicomachean Ethics*. Trans. H. Rackham. In *Aristotle: The Nicomachean Ethics*. Loeb Classical Library, 1934.

——. *On the Soul*. Trans. W. S. Hett. In *Aristotle: On the Soul; Parva Naturalia; On Breath*. Loeb Classical Library, 1957.

Bohm, D. 1957. *Causality and Chance in Modern Physics*. London: Routledge and Kegan Paul.

Burtt, E. A. 1932. *The Metaphysical Foundations of Modern Physical Science*. Revised ed. New York: Harcourt, Brace.

Crick, F. 1995. *The Astonishing Hypothesis: The Scientific Search for the Soul*. New York: Scribner.

Dawkins, R. 1986. *The Blind Watchmaker*. New York: W. W. Norton.

Deferrari, R. J., et al. 1948. *A Lexicon of St. Thomas Aquinas*, based on the *Summa Theologica* and selected passages of his other works. Washington, D.C.: Catholic University of America Press.

Descartes, René. 1641/1990. *Meditationes de prima Philosophia/ Meditations on First Philosophy*. Bilingual edition. Ed. and trans. G. Heffernan. Notre Dame, Ind.: University of Notre Dame Press.

——. 1991. *The Philosophical Writings of Descartes*. 3 vols. Trans. J. Cottingham, R. Stoothoff, D. Murdoch, and A. Kenny. Cambridge: Cambridge University Press. (Vols. 1 and 2 first published 1985 and 1984 respectively.)

Dupré, J. 1993. *The Disorder of Things: Metaphysical Foundations of the Diversity of Science*. Cambridge: Harvard University Press.

Fisher, R. A. 1930. *The Genetical Theory of Natural Selection*. Oxford: Clarendon Press.

Freeman, K. 1957. *Ancilla to the Pre-Socratic Philosophers*. Cambridge: Harvard University Press.

Grünbaum, A. 1967. The status of temporal becoming. In *The Philosophy of Time*, ed. R. M. Gale, 322–53. New York: Doubleday.

Hahn, R. 1967. Laplace as a Newtonian scientist. A paper delivered at a seminar on the Newtonian influence held at the Clark Library, 8 April 1967. Los Angeles: William Andrews Clark Memorial Library, University of California, Los Angeles.

Held, B. S. 1995. *Back to Reality: A Critique of Postmodern Theory in Psychotherapy.* New York: W. W. Norton.

Hume, D. 1748/1955. *An Inquiry Concerning Human Understanding.* Ed. C. W. Hendel. Indianapolis: Bobbs-Merrill (Library of Liberal Arts).

Kant, I. 1781/1787/1929. *Critique of Pure Reason.* Trans. N. K. Smith. London: Macmillan. (A translation of the text of both the first [A] and the second [B] editions of Kant's *Kritik der reinen Vernunft.*) Paperback reprints 1965 and various dates after.

Laplace, P.-S. 1814/1921. *Essai philosophique sur les probabilités.* Ed. M. Solovine. Paris: Gauthier-Villars. [*A Philosophical Essay on Probabilities.* Trans. F. W. Truscott and F. L. Emory. New York: J. Wiley, 1902. Reprint, New York: Dover, 1951. This translation was not used in this book: the passage quoted in Chapter 3 was translated by the author.]

Leibniz, G. W. 1710/1952/1966. *Theodicy: Essays on the Goodness of God, the Freedom of Man, and the Origin of Evil.* Ed. A. Farrer, trans. E. M. Huggard. London: Routledge & Kegan Paul, 1952. *Theodicy* (abridged). Ed. D. Allen. Indianapolis: Bobbs-Merrill (Library of Liberal Arts), 1966.

——. 1714/1898. *The Monadology and Other Philosophical Writings.* Trans. with intro. and notes by R. Latta. Oxford: Oxford University Press. Reprints various dates.

Locke, J. 1690/1975. *An Essay Concerning Human Understanding,* Ed. with intro., critical apparatus, and glossary by P. H. Nidditch. Oxford: Clarendon Press.

Nagel, E. 1961. *The Structure of Science.* New York: Harcourt, Brace & World.

Plato. Works of Plato are cited in the text by the standard (Stephanus) pagination, which usually appears in the margins of good modern editions. The following editions of certain of Plato's works were used by the author but are not cited in the text.

——. *Phaedo.* Trans. H. N. Fowler. In *Plato: Euthypro, Apology, Crito, Phaedo, Phaedrus.* Loeb Classical Library, 1914.

——. *Phaedo.* Trans. with intro. and commentary by R. Hackforth. Cambridge: Cambridge University Press, 1955. Paperback reprint, Indianapolis: Bobbs-Merrill (Library of Liberal Arts), various dates.

——. *Phaedrus.* Ibid.

——. *Phaedrus.* Trans. R. Hackforth. In *Plato's Phaedrus:* With introduction and commentary. Cambridge: University Press, 1952.

——. *Philebus.* Trans. H. N. Fowler. In *Plato: Statesman, Philebus, Ion.* Trans. Fowler and W. R. M. Lamb. Loeb Classical Library, 1939.

——. *Sophist.* In *Plato's Theory of Knowledge:* The *Theaetetus* and the *Sophist* of Plato trans. with running commentary by F. M. Cornford. London: Kegan Paul, Trench, Trubner, 1935.

——. *Timaeus.* Trans. R. G. Bury. In *Plato: Timaeus, Critias, Cleitophon, Menexenus, Epistles.* Loeb Classical Library, 1929.

——. *Timaeus.* In *Plato's Cosmology: The Timaeus of Plato.* Trans. with running commentary by F. M. Cornford. London: Routledge & Kegan Paul, 1937. Paperback reprint, Indianapolis: Bobbs-Merrill (Library of Liberal Arts), various dates.

Pols, E. 1963. *The Recognition of Reason.* Carbondale: Southern Illinois University Press.

——. 1975. *Meditation on a Prisoner: Towards Understanding Action and Mind.* Carbondale: Southern Illinois University Press.

——. 1982. *The Acts of Our Being: A Reflection on Agency and Responsibility.* Amherst: University of Massachusetts Press.

——. 1992. *Radical Realism: Direct Knowing in Science and Philosophy.* Ithaca: Cornell University Press.

Russell, B. 1912. On the notion of cause. *Proceedings of the Aristotelian Society*, 1912–13, 1–26. Reprinted in Russell, *Mysticism and Logic*. London: Longmans, Green, 1918, 180–228.

Searle, J. R. 1992. *The Rediscovery of the Mind*. Cambridge: MIT Press.

Simpson, G. G. 1953. *The Major Features of Evolution*. New York: Columbia University Press.

Suppe, F. 1974. The search for philosophic understanding of scientific theories. In *The Structure of Scientific Theories*, ed. Suppe, 6–232. Urbana: University of Illinois Press.

Wheelwright, P. 1966. *The Presocratics*. New York: Odyssey Press. Paperback reprint, various dates.

Index

DATE DUE

DEMCO 13829810